THE
BEAR
AT THE
BIRD
FEEDER

THE BEAR AT THE BIRD FEEDER

WHY WE'RE SEEING MORE WILD ANIMALS IN OUR NEIGHBORHOODS AND HOW WE CAN LIVE IN HARMONY WITH THEM

RANDI MINETOR

LYONS PRESS

Essex, Connecticut

An imprint of The Globe Pequot Publishing Group, Inc.
64 South Main St.
Essex, CT 06426
www.GlobePequot.com

Copyright © 2025 by The Globe Pequot Publishing Group, Inc.

All rights reserved. No part of this book may be reproduced in any form or by any electronic or mechanical means, including information storage and retrieval systems, without written permission from the publisher, except by a reviewer who may quote passages in a review.

British Library Cataloguing in Publication Information available

Library of Congress Cataloging-in-Publication Data available

ISBN 9781493089499 (paperback) | ISBN 9781493089505 (epub)

*To Steve Nicandri, my seventh-grade science teacher,
for using* Silent Spring *by Rachel Carson as a textbook,
introducing me to the endangered creatures
that share our world and the need to protect them.*

Contents

Introduction 1

Chapter 1: Please Don't Feed the Bears 7
Chapter 2: Secretive Neighbors: Mountain Lions 29
Chapter 3: Fiercely Independent: Bobcats 41
Chapter 4: Back with a Vengeance: Wild Turkeys 51
Chapter 5: So. Many. Deer. 65
Chapter 6: At Home with Bats 79
Chapter 7: Top of the Food Chain: Coyotes 93
Chapter 8: The Unexpected Moose 109
Chapter 9: Squirrels: Captains of Industry 121
Chapter 10: No Goose, No Goose . . . Giant Goose!
  The Canada Goose Returns 133
Chapter 11: Thieves in the Night: Raccoons 145
Chapter 12: As Clever as They Look: Foxes 157
Chapter 13: Assume There Are Alligators 169
Chapter 14: Fragrant Neighbors: Skunks 181

A Final Word 195
Acknowledgments 197
Selected Bibliography 199
Index 213

INTRODUCTION

THE PHOTO WENT VIRAL ALMOST BEFORE PHOTOGRAPHER MICHAEL Novo knew what was happening: an image of a lone coyote trotting down Michigan Avenue in the heart of Chicago, Illinois, as the 2020 COVID-19 pandemic lockdown got underway. With the streets absolutely empty of people and the traffic lights glowing green, it seemed as if this wild animal sensed the absence of humanity from afar, coming out of the Midwestern plains to take possession of this usually congested intersection. Those of us old enough to remember the rash of post-apocalyptic science fiction films of the 1960s recalled the creepy feeling we got from *The World, the Flesh and the Devil* or *On the Beach*, seeing such tall buildings and normally congested streets with one lonely soul still in motion.

Novo posted this photo "in just one sub on Reddit and that's where it all began," he wrote on his personal website. In minutes people began sharing the image across their social media, and soon he found himself inundated with licensing opportunities from broadcast outlets, including major national networks. The image illustrated an eerie natural phenomenon: the animals stood ready to take back our cities. Soon videos of animals walking down city streets sprang up all over the internet, from coyotes in the middle of San Francisco to Nubian ibexes in Eilat, Israel; wild mountain goats scampering along sidewalks in Llandudno, Wales; a puma exploring the roads in Santiago, the capital of Chile; raccoons parading down paved paths usually filled with joggers, walkers, and tourists in New York's Central Park; and red foxes and their kits playing on sidewalks in Toronto, Ontario.

Were the world's wildernesses about to empty their wild residents into our cities? How did the animals know to come so far and take back the streets?

Here's how: they were already here.

For centuries, humans have laid claim to open land and turned it into housing developments, industrial sites, office space, and pavement, creating places in which we can live and work. Open plains and meadows became skyscrapers and factories, or farm fields full of crops and livestock. Forests fell and mountains lost their coniferous covering, their trees milled into lumber to build homes and businesses. As the consequences of these actions came to light, some local and state governments have taken steps to restore what they took from the landscape, planting new forests to prevent erosion of entire mountainsides and requiring developers to preserve some land as forever wild, as a trade-off for the property they acquire. In my home state of New York, two large land masses have become Adirondack Park and Catskill Park, models of preservation for the entire country to emulate: permanently preserving millions of acres of mountainous forest alongside developed areas, for the good of the environment and its inhabitants.

As we usurped so much land for our own purposes, it was easy to assume that all of the wildlife that lived there had either moved on to find a new home or simply died. Then some wild animals came to prey on what humans could offer them: calves and lambs on the edges of herds, plump chickens in farmyards, delicious fruits and vegetables in fields or backyard gardens, food we put out for birds, even small pets. Had predators like black bears, mountain lions, bobcats, or foxes come out of faraway forests to plunder our towns and yards? The answer is much closer to home: these stealthy creatures learned to skulk around the edges of human settlements and watch for the right opening to approach us. They have remained right where they have always been, watching as we took over the land where they had always lived, and they found ways to adapt to our presence.

In many cases, this prolonged proximity to humans put the animals in real peril. Some of them became favorites of the fur trade, teetering on the edge of extinction as trappers gathered every pelt they could get their hands on. Some animals paid the price for preying on livestock, with

Introduction

bounty hunters bringing in thousands of dead predators to keep them from plundering cattle and sheep in pastures. Some nearly vanished in an era when there were no limits to the number of geese, turkeys, or deer that hunters could take in a season—because every season was hunting season. And some were spared when humans discovered ways that these animals were actually more useful to them alive than dead.

Some of the animals driven nearly to extinction had the opportunity to make a comeback. Seeing the animals dwindle, state and local governments made changes to hunting laws, restricting them to times of year when taking the adults would not orphan their nursing young. Some recovered when the fickleness of fashion trends ended the demand for fur and animal hides. The onset of the conservation movement in the 1970s funded the work of organizations that preserve open spaces for the benefit of humans as well as wildlife, and passage of the Endangered Species Act in 1973 protected many animals across the country.

Over time, as the human footprint grew and the only choice remaining to animals was living among us, many of them saw advantages to having human neighbors. Careless humans don't always lock up their garbage at the end of the day. We leave bowls of pet food on porches and in backyards for our dog's or cat's convenience. We build structures like sheds, decks, and porches that sit right on the ground, making it easy for animals to dig themselves cozy dens and burrows underneath them. Our presence even keeps larger predators away, making it safer for smaller creatures to raise their young.

The wild animals in our midst have bided their time, learning which of their behaviors encourage humans to supply the food, shelter, and protection they need, and which ones trigger us to become threats to their existence. Some have profited from being adorable, delighting people with their young's antics or making every observation a special treat. Others have pushed the boundaries of what their human neighbors will tolerate. And a few have become truly dangerous.

This wide range of relationships between humans and wildlife has made the balance between animal needs and human needs difficult to maintain. It gets even harder when so many humans have different opinions on what this balance should look like.

Let's take a random cul-de-sac in a suburb somewhere in America, where the homes have backyards that end in a strip of woodland. Here, a white-tailed doe has quietly given birth to a fawn just inside the woods, under a pile of blown-down branches and brush. In a few days, the doe and fawn wander into the nearest yard, and the doe begins munching on the leaves and twigs of the homeowners' shrubs. One homeowner is delighted to see this new fawn right in his own backyard, so the family pulls chairs up to the bay window to watch as the deer explore their property.

The neighbor to the left of the homeowner comes out on her deck to watch as well, and soon the two homes are texting daily observations about the fawn, how much it has grown, when it is likely to stop nursing and start feeding on the shrubs along the back of the property, and so on. The neighbor on the right, who works from home and has a constant view of the backyard, does not join in with the others. He feels no delight in having deer in the yard, so he bangs out of his back door several times a day to scare the doe and fawn away from his bushes. After several days of this, he calls the municipal animal control office to report a "deer problem."

When an animal control staff member arrives to determine the extent of the problem, she finds the doe and her fawn calmly eating from another neighbor's shrubs. The other neighbors, she soon determines, are shocked that someone considers this bucolic situation a problem at all. Animal control recommends that if the man does not want deer in his yard, he should put up a fence to keep them out. He argues that he should not have to take on that expense, so the deer should be relocated—a process only undertaken with animals that have become so habituated to people that they present a hazard to themselves and the people around them.

If this sounds familiar, you can see how difficult it is to resolve a situation involving animals and people to everyone's satisfaction. Not only is one neighbor's problem another neighbor's pleasure, but some may not be willing to take on the responsibility of managing their own issues and property. The animal quickly becomes the cherished experience of one and the enemy of the other, just by doing what it does as a natural part of its lifestyle.

Peaceful coexistence is possible when humans acknowledge their personal role in achieving it. It may take some effort, some important reminders, a few potentially inconvenient precautions, and an investment in creating boundaries between us and the animals, but we can share our world safely and comfortably with the creatures around us.

The Bear at the Bird Feeder takes a close look at fourteen animals that have become common in or near our cities and towns, with the goal of providing insights readers can use to live in harmony with these wild creatures. In each chapter, I provide the best advice I have gathered from wildlife professionals, researchers, and other sources to help you keep yourself, your family, and your property safe from the occasional mishaps that happen when wild animals pay a visit. I have interviewed people who have had positive encounters with one or more of these critters, and I have also searched for stories of people who do not take every precaution with the animals in their midst.

Not everyone was willing to talk with me. In fact, I was surprised at the number of government agencies, wildlife organizations, and viral video makers who did not return my messages or who sent me off on endless cycles of handoffs, only to reach an eventual dead end. It would not have occurred to me before I started this project that topics like wild turkeys on a city street, bats in an attic, or feeding raccoons off the back porch would rise to the level of such controversy, but they certainly have. Occasionally, I have used incidents from my own experience to fill in the gaps in expert commentary.

What you will find in this book is the result of extensive research to determine the best courses of action in a wide range of scenarios, as well as some examples of what works and what does not when animals decide to make your property their own. In the end, your relationship with the animals in your neighborhood is a highly personal choice, one that I hope you will not make lightly. If this book helps you make more informed choices of what to do in your own home and backyard, then I have done my job.

No matter how placid it may look, an American black bear should make people uneasy if it decides to frequent their backyards. @ KYLE KEMPF, ISTOCK

Chapter 1

PLEASE DON'T FEED THE BEARS

The residents of David Oppenheimer's neighborhood in suburban Asheville, North Carolina, have come to consider American black bears (*Ursus americanus*) part of the ample scenery. The area's hardwood forests include thousands of acres of mountain oaks, producing bounteous piles of acorns from late summer through the fall. This alone offers the local bears plenty of natural reasons to frequent the neighborhood's yards and open lots, as David noted when we spoke via Zoom on a mid-June afternoon. But there's even more to attract the bears in Asheville.

"Beyond the acorns, we have blueberry and blackberry bushes in the open areas between houses," he said. "I'll walk by and see bears standing up on their hind legs, eating all those berries."

Asheville sits at a crossroads between the Pisgah and Nantahala National Forests to the northeast and southwest, Cherokee National Forest to the northwest, and Great Smoky Mountains National Park just an hour's drive west—where nearly two thousand bears make their home, equating to about two bears per square mile. With this nearly complete surround of traditional bear habitat, it surprises no one that the bears frequent this lushly forested housing development and others in the Asheville area.

"You see people out walking in the evening and they stop at one of the open areas and watch the bears eating," David said. "Some people don't like them—they feel like they're trapped in their homes—but most people just accept that they're there."

The bears seem to enjoy an understanding with their human neighbors: neither species intends to prey on the other. "Now they walk down the street in the middle of the day, lay down in a yard and take naps," he noted. "The bears are not stalking or seeking out people. Some people yell at them, and other people just love them. You see little old ladies walk right past them with their poodles."

Generally, the bears reciprocate by leaving the neighbors alone. The temptation of human food aromas often becomes too great to ignore, however, especially during late summer and fall when adult bears begin to pack on pounds of fat to sustain them during their long winter hibernation. As delicious as wild berries are, a pound of berries provides a bear with less than three hundred calories, while half a cup of discarded bacon grease discovered in a trash can, for example, supplies nearly one thousand calories with considerably less effort. These comparison statistics and many more come via Linda Masterson, communications director for the national education and outreach program BearWise™, of the Association of Fish and Wildlife Agencies, and author of the comprehensive *Living with Bears Handbook*.

"Bears are always trying to gain weight," she said in a Zoom interview. "So the more food they get for the least amount of work, the better. They will explore anything that smells like it might be food. It's eat or die."

As much as 90 percent of a bear's diet is plant-based, but bears are true omnivores, seeking out insects, fish, and carrion in spring when plant matter is scarce, and again in late summer and fall as they bulk up ahead of the long winter. Before their hibernation, bears enter a state known as hyperphagia—insatiable hunger that drives them to eat everything they can find that even remotely resembles food. When they deplete berry and plant supplies and even dead animals left behind by hunters, this critical time can drive black bears to seek out the contents of human garbage dumps, trash cans, bird feeders, and grills, and treats like pet food left outside for cats and dogs.

David and his neighbors have taken every precaution to keep their household garbage out of bears' reach, keeping their trash in the freezer or locked in the garage with all the doors secured. Most of the residents have invested in bear-proof trash cans that make it virtually impossible for bears to get at what's inside. "The bears can beat on it all day, but they're not going to get any opening split," he said. "You can see all the bear claw marks on them, but they don't get in." The locked cans create a small inconvenience, he acknowledged: they have to stay locked right up until the trash collectors come to empty them. "One of the tricks is getting the can out on trash day; I can't put it out and unlock it the night before. So I have to go out early on trash day to open the lock."

But David sees this as a small price to pay for the privilege of sharing this neighborhood with the fascinating animals. He invested in a bear-proof bird feeder, a solid metal contraption that only dispenses seed through small holes fit for finches and other songbirds. Suspended from a complicated pulley system that keeps them high in a tree, the feeders stay out of reach of bears on the ground—but black bears are agile tree climbers, managing to reach the feeders and quickly figuring out how to turn them over until the seed pours out of the tiny holes. A pound of black oil sunflower seed can provide more than 2,500 calories, Masterson notes in her book, making the effort of investigating the feeder and deducing a solution absolutely worth it to the bear.

"So I stopped feeding the birds in spring and summer, when the bears are active," David said. "Now I just feed them in the winter, when the bears are asleep."

It seemed as if David and his neighbors have got their lives with bears pretty well figured out—until the day that a Ring doorbell camera captured a potential life-and-death moment that became a viral sensation.

A commercial photographer with a specialty in rock concerts and other live events, David's work can be physically demanding—so like most of us at the end of the day, he enjoys some down time. On this pleasant afternoon in mid-April 2023, he sat back in a chaise lounge chair on his carport, relaxing with the news and correspondence on his phone. He barely looked up when his Ring app jingled to let him know that someone had passed close to his door. He glanced at the Ring app and did not see any bears, so he thought that the alert could just be a neighbor walking a dog, or a squirrel, or even a bush rustling in the breeze. He continued scrolling until the visitor ambled into his view.

A young black bear stood barely eight feet from him.

David started, his eyes growing wide as he froze in place. In an instant he realized the vulnerability of his supine position. He covered his midsection with the only defense he had handy: a small, white throw pillow.

The bear did not notice him immediately, but his audible gasp alerted it, startling the animal into motionless silence.

For a tense four seconds, the two locked eyes on one another.

"You're not supposed to stare down a bear, but I didn't want to turn my head," David told me. "I do have bear spray, but not there with me. If I was hiking in Colorado, it would be different, but this was at home."

David knew what the recommended actions were for encountering a bear at close range: stand tall, try to look big, wave your arms over your head, and shout. But in those precious seconds, he waited without moving a muscle, giving the bear the option of making whatever first move it chose.

The bear chose well. When David moved the white pillow slightly, it seemed to break the stoic bear out of its trance-like stare. It turned around and ran off.

David began to breathe again. He'd had a close call, but in retrospect, he doesn't believe he was in any danger.

"Bears are smart, they're intuitive," he said. "They can perceive a threat. Obviously, they're wild animals, so you don't want to push your luck. But this one seemed to know I wasn't going to hurt it."

About six weeks later, as he sat comfortably on his outdoor loveseat once again, a black bear poked its nose out from behind David's parked car and made its silent way to his outstretched feet. This time David didn't hesitate. He leapt up and had the presence of mind to grab his phone and start recording the visit. His sudden move scared the bear back behind the car, however, so David began attempting to console it.

"Hi bear, I didn't mean to scare you," he crooned. He backed off to a spot just beyond the carport, until the bear popped up from behind the vehicle to look at him.

"It's okay, you can walk by," David coaxed, and the bear seemed to take him at his word. As David captured video for another two minutes, the bear made a thorough exploration of the carport and front porch, passing within two or three feet of the man whose leavings it sought to plunder. Despite its careful canvass, however, the bear found nothing to eat, so it finally made its way off the porch and lumbered away across the yard.

Were these close calls, or simply visits from curious animals? David believes that neither of the bears meant him any harm, and he has seldom seen one of the neighborhood's bears so much as pretend to charge at a human being. Nonetheless, he has made one change on his property since these incidents.

"I turned my chair around so I can see what's coming while I'm out there," he said.

Black bear cubs see backyards and porches as places to play. @ DUSTIN SAFRANEK, ISTOCK

The Wild and the Innocent

Linda Masterson loves the outdoors, so much so that she walked away from the obligations of corporate America in 2001 to spend more time in one of America's most spectacular areas. She and her husband moved from Chicago to the Rocky Mountains, where she joined the Colorado Division of Wildlife's Bear Aware team as a lead volunteer, while running her own marketing and communications firm.

"I have always been very interested in people's relationships with nature and wanting to help people do a better job with that, even when I was in marketing and advertising in Chicago," she said. "We moved from Chicago in the middle of a very bad year for bears—we didn't know it at the time. My late husband and I joined the Colorado Division of Wildlife, which was just launching what was then their Bear Aware program."

Bear Aware aimed to teach people how to do a better job of coexisting with wildlife, with a sharp focus on bears, she said. "It was a massive food failure year, and there were a lot of issues with bears coming into town, breaking into vehicles, breaking into remote cabins and helping themselves to all the stuff people left there. They really wanted to educate the public. I was a lead volunteer with them for eleven years."

Since then, she has immersed herself in communications activities that teach people about the bears in and around their neighborhoods, camps, and favorite wilderness areas. Working closely with noted Colorado bear biologist Tom Beck, an association that opened doors to a wide range of resources, Linda wrote the first edition of her *Living with Bears Handbook* in 2005 and worked with Washington's bear specialist Rich Beausoleil on the recently updated and expanded second edition. Today she serves on the organizing committee for the International Human–Bear Conflicts Workshop, and she's a member of the International Association for Bear Research and Management. She's responsible for communications for BearWise, a program supported by most state wildlife management agencies and dedicated to helping people live responsibly with black bears. To date, state-level wildlife organizations in forty-one states have become BearWise members.

In our June 2024 Zoom conversation, and in subsequent emails, Linda noted that while some states have fewer than one thousand black bears roaming their wild areas, more than forty of the contiguous forty-eight states have some measurable population of them—and their numbers are growing. "A state-by-state comparison shows that the human population in the Eastern US grew by 5.8 percent [from 2015 to 2022]; the bear population grew by 3.9 percent," she said. "And just to clarify, bears don't fill out census forms. Some states have very sophisticated and complex methods of estimating populations, and some are making educated estimates."

This recent good news for bears came after hundreds of years of very bad news, beginning about two centuries ago with the arrival of millions of Europeans on North American soil and the growth of every United States industry. In the Eastern states, millions of acres of forest fell in massive lumber harvests, feeding the construction of homes, barns, factories, and churches, and supplying wood pulp for the paper industry. Builders cleared the land and laid out cities and industrial centers, pulling out stumps and ensuring that no forests would ever grow in these areas again.

"In the late 1800s, wildlife was either a commodity to be used, or a problem to be exterminated," wrote Masterson in *Living with Bears Handbook*. "There were no hunting seasons or rules and regulations; bears were plentiful, and bounties so common hunters could make a good living selling bear pelts."

Meanwhile, the discovery of rich deposits of minerals deep within the Western territories' mountains upended bear populations even further. Mining practices often involved clear-cutting woodlands to make room to operate heavy machinery, house thousands of mine workers in pop-up towns, and transport the unearthed materials to processing facilities.

By the early twentieth century, bears had virtually disappeared in several states—and not just black bears. Grizzly bears also became scarce, seen only in areas protected from development by the newly founded US Forest Service. In 1916, the fledgling National Park Service joined the effort to provide forever habitat for grizzly and black bears, as well as many other woodland animals. Unfortunately, it was already too late for most grizzlies: despite gaining significant status as the most prominent emblem on the state flag of California, grizzly bears succumbed to persecution and hunting in that state, disappearing entirely from California by 1924. (California named the grizzly bear its state animal in 1953, nearly thirty years after the bear had been driven to extinction there.)

Bears across the continent found themselves hemmed in by human development and limited by the amount of available habitat—but none of this prevented them from continuing their annual life cycle of waking from hibernation, raising their young, satisfying their voracious hunger, and returning to the den for another winter. When trees and shrubs did not sate their appetites and pile on the fat they needed to sustain them through hibernation, they looked to their new human neighbors to provide more satisfying meals. Black bears appeared near farms and around the outskirts of settled areas—and farmers and others, seeing the large animals as threats to their livestock and perhaps to their own lives, declared open season on them. By the mid-1900s, black bears had virtually disappeared from many American landscapes.

Then in the late 1950s and early 1960s, scientists all over the world began calling for change. Rachel Carson's landmark book *Silent Spring*, published in 1962, revealed the destructive power of the pesticide DDT, which not only affected humans but had driven bald eagle and other raptor numbers to near extinction. When Richard Nixon became president in 1969, he began working with Congress to develop environmental laws—and the first of several important bills to come out of this effort was the Endangered Species Conservation Act, expanding the list of animals that would receive federal protection under the law.

Black bears nationwide never landed on the endangered species list, but their dwindling populations in Florida and Louisiana did. The US Fish and Wildlife Service (USFWS) listed the Louisiana bears as "threatened" in 1992, triggering a cascade of federal, state, and local efforts to restore the bears' habitat and restrict hunting of them. The Louisiana Wildlife and Fisheries Foundation reports that it has restored or protected more than 750,000 acres of bear habitat since 1992, creating a sustainable black bear population that is expected to maintain itself throughout the twenty-first century. In May 2016, USFWS delisted Louisiana's black bears, a major triumph for the state's preservation efforts.

Today, black bear numbers across the continent have rebounded with vigor, a testament to the effects of habitat preservation efforts like Louisiana's program. Total estimated populations roaming North America exceeded 950,000 black bears in 2015, including very healthy numbers in Canada, Masterson said. The Eastern United States is seeing significant growth, with the heavily forested states of Virginia, West Virginia, New York, North Carolina, and Tennessee supplying many miles of contiguous forests. In the West, Colorado, Oregon, and Washington each host populations well over 10,000, while Minnesota and Wisconsin's northern forests provide habitat to roughly 15,000 and 24,000 black bears, respectively.

Don't Call It a "Bear Problem"

Considering the growth in bear populations across the continent, the number and frequency of bear–human interactions is surprisingly small. BearVault, a Colorado company that makes bear-resistant canisters, reports that since records were first kept in 1784, 180 fatal bear/human conflicts have taken place—and only sixty-six of these involved wild black bears (that's one every three-plus years). In most years, fewer than a dozen nonfatal physical encounters happen between people and black bears, "because black bears are far more likely to run away from you than engage," writes Jessica Cockcroft in a 2023 blog on the BearVault site.

Bear biologist Lynn Rogers, PhD, of the North American Bear Center, notes that in more than fifty years of working with black bears, he has observed enough of their behavior to dispel the many myths that form an ominous mystique about these large, powerful creatures. He has endured a slap from a bear that scratched welts into his torso, he writes, but "the damage from a slap is nothing close to the folklore that a bear can disembowel man or beast with a swipe of the paw. Black bear claws are strong for climbing trees, but not sharp for holding prey."

He also dismisses the assumption long held by campers and hikers that coming between a mother bear and her cubs can trigger an aggressive attack. "A big revelation to me was how reluctant black bear mothers are to defend their cubs against people, even when the family is cornered in a den and I'm trying to stick the mother with a needle to tranquilize her," he writes. "There is no record of anyone being killed by a mother black bear defending her cubs, and attacks are very rare. We routinely capture black bear cubs in the presence of mothers and have never been attacked."

Stephen Herrero, widely considered the foremost expert in North America on bear attacks and the author of *Bear Attacks: Their Causes and Avoidance* (now in its third edition), does his best to calm the trepidation of people who will not venture into the wilderness for fear that they will be attacked and even eaten by a bear. "Each year there are millions of times in which each species [of bear] is close to people and no threat or injury results," he writes. "In Canada and the United States during the decade of the 1990s, bears killed twenty-nine people; grizzly bears killed eighteen; and black bears, eleven. To put this into some perspective . . . between 1977 and 1998, dogs killed 250 North Americans."

On the other hand, bears can become habituated to people, simply by having repeated encounters with us from a distance. "A new noise, sight, or smell usually catches an animal's attention, and it responds in some way," he writes. "But if the stimulus is presented repeatedly and nothing good or bad happens, then the animal becomes used to the stimulus, and attention and response wane. The animal has become habituated to the stimulus." This is what happens when bears discover that people frequent their accustomed habitat, whether these people are hiking on trails in a wilderness area, walking their dogs in a neighborhood, or lounging in their own carports at their homes.

"I think there are bears out there wondering, 'Why are all these people in our backyard?'" said Linda Masterson in our conversation. "It's a combination of way more people in bear country—today more people than ever before are living and playing in bear country. Thousands of acres of natural habitat are lost every year."

This perspective turns the issue on its ear. It's not that bears are moving into human habitat; instead, we are moving into bear habitat. "Bears live in places that people find very desirable," she said. "And bears need connected corridors," ways to move between one habitat and the next in their search for food. "Young male bears can travel long distances looking for a place of their own. Mature male bears roam around looking for mates. There are a lot of bears moving on the landscape. Sometimes your new housing development is in the middle of what was a connected corridor, an area the bears had moved through for generations."

And as the bears travel through, they discover a compelling reason to hang around: significant quantities of delicious, high-calorie food. "The entire time bears are not hibernating, they're eating," said Masterson. "Mom has to travel afield to find food for cubs, and to teach them to find food for themselves. If she teaches them to raid dumpsters and to go through doggie doors into houses, they will remember, and they will go back to places that provided them with a great meal."

Bears can live for as long as thirty years, so what they learn to eat as cubs can become part of a lifelong diet, whether it's a berry patch, a dog food from a bowl on a porch, or trash in a dumpster left unlocked and unattended.

This learned behavior is what Herrero calls "food conditioning." The bear becomes increasingly comfortable with food sources in relatively close proximity to people and eventually forms an association between people and food. "The food-conditioned bear is almost always somewhat habituated to the smell or sight of people," he notes. This bear learns to plunder a loosely covered trash can to get at morsels discarded by human residents, or to tip over a bird feeder to pour out several pounds of sunflower seeds. Over time, the bear's highly developed sense of smell may lead it to crawl under a partially open garage door, slip through a dog door, break a window, knock a lid off a grill, and even learn to manipulate an unlocked car door to get at the satisfying treats inside.

Bears that choose not to resist these temptations quickly develop a reputation for being "problem" bears, a term that Masterson is quick to disallow. It's easy to blame the bear when the people in its habitat supply a constant smorgasbord of readily acquired calories, she points out, but the issue becomes compounded when homeowners dismiss food-conditioned bears as "someone else's problem." Just one careless neighbor can offer enough accessible and unprotected trash, bird seed, grill leavings, pet food, or livestock feed to bring a whole family of bears into regular contact with the neighbor's home.

"If you choose to live in the bears' backyard, it's up to you to learn how to do so responsibly," she says in her book. "Bears have proved again and again that they can adapt to our presence ... without disrupting the daily lives of their human neighbors. Unfortunately, people often prove to be far less adaptable than bears."

Once a bear's pattern of raiding a home's garbage or bird feeders has been established, it can be very hard to break. Homeowners may assume that their state's wildlife management professionals will hop in a truck and come out to trap and relocate a food-conditioned bear. Most wildlife agencies agree that this only happens as a last resort, however, and only if the bear shows "undesirable behaviors in the presence of humans," as detailed by the *New York State Black Bear Response Manual*, a publication of the state's Department of Environmental Conservation (NYSDEC). New York's Adirondack Park and Catskill Park, two of the largest publicly protected areas in the contiguous United States, have significant populations of black bears living in close proximity to people's homes. Wildlife management practices in rural New York state have become a model for many other states to follow.

Undesirable bear behaviors, NYSDEC explains, include habitual visits to a human-supplied food source, a lack of fear of humans, and a lack of response to "aversive conditioning techniques" like scare darts, rubber buckshot, and loud noisemakers. These bears "have not necessarily shown aggression toward people," but they have lost their fear of humans and must be removed from their familiar sources of food in hopes that they will change their habits.

Trap-and-relocate operations usually involve several different agencies in a costly and time-consuming process—and the effort may turn out to be for naught, NYSDEC says. "A relocated bear will often travel great distances to return to where it was originally captured," its policy says. "If it can't find its way back, it will often seek out new human food sources in the area where it was released." Zoos and wildlife sanctuaries already have all the black bears they need, so life in captivity is not an option. Repeat offenders may have to be euthanized.

When It's a People Problem

Search "people feeding bears" on YouTube, and you'll find dozens of videos of people deliberately changing the behavior of wild bears by offering them treats. At a lodge in Gatlinburg, Tennessee, just outside of Great Smoky Mountains National Park, guests toss food to waiting bears and their cubs in the Pigeon Riverbed below them. A man in northern Minnesota began feeding one bear during a year when natural food was scarce, and now has as many as ninety—yes, ninety—black bears frequenting his property, turning it into a bear viewing area for tourists. A video that begins with a camouflage-clad man grinning and saying, "Let's give the bears some treats!" shows him hand-feeding several bears who are obviously conditioned to expect this and even do tricks for him. Hundreds of thousands of likes on these and many other videos encourage this behavior, even though most videos are accompanied by narration that reiterates all of the reasons that feeding bears can endanger both bears and humans.

Some communities are working to curb this kind of activity. In Asheville, North Carolina, where David Oppenheimer lives, the state Wildlife Resource Commission reports that it receives calls every day about people feeding bears—not just bears plundering trash cans or bird feeders, but actual piles of food placed outside a window or porch, where bears devour it for the entertainment of the human residents. North Carolina special projects biologist Ashley Hobbs told WLOS-TV in May 2024 that people either fill bird feeders specifically for the bears, or they put candy bars or heaps of nuts in their backyards to attract them.

"Feeding bears is turning into property damage, either for that person or their neighbors in that community, or it's creating unsafe situations for the people living in that community," she said.

Recognizing that these unwanted actions might be curtailed by consequences, the commission worked with the city and county to create a bear-feeding ordinance. This local legislation says that placing food out for bears can result in a fine—and if people become repeat offenders, the fine gets larger: it may start at fifty dollars but repeat offenders could find themselves fined up to five hundred dollars. If a resident refuses to pay the fine, the city can put a lien on their home—and at the county level, ignoring the ordinance can be a criminal offense.

"We get a lot of human–bear interaction calls," Hobbs told me in a Zoom call. "Someone saw a bear on their Ring camera, there's a bear in the kitchen—these can be related back to unsecured attractants," like pet food left out on a porch, or people food on a counter with nothing but a screen door between the food and a bear. "But we also get people feeding bears turkey sandwiches off the back porch. This ordinance lets us respond to that immediately, and make those situations stop."

One of Asheville's residents loves seeing bears so much that he buys them doughnuts and pastries at local grocery stores, dumping these treats along Town Mountain Road, a street that passes through housing developments in suburban Asheville. The bears collect by the roadside to feast on these processed foods. "He thinks he's helping the bears, but he's actually drawing them down there," said Hobbs. Bears congregating on the road's shoulders are in danger of collisions with cars, especially as they become accustomed to associating people with food. "I got reports of this from the state agency, and I turned it over to county animal control. They are enforcing the ordinance."

Whether the fines are working is hard to say, said Hobbs, as local governments may not have the staff and resources to keep up with the number of cases. "But just knowing that a fine is a possibility, maybe they hear about the ordinance, and they change their tune," she said. "I hope people understand why we're doing this, and how it helps the bears. It protects public safety, property, and the bears themselves."

It's altogether too easy to regard the neighborhood bears the way they have been portrayed in old Disney movies or modern-day animated sitcoms, but they are truly wild animals, said Hobbs. "Bears are reflections of us," she said. "They are responding to our behavior and our activities. We are more or less in complete control of bears in our community—if there's nothing for them to eat, they won't hang out in the yard. If we back up when we see them and give them space, they will do the same."

What You Can Do

The best solution to living in bear country is to avoid the issue altogether. Bear biologists, researchers, and wildlife management organizations across the country recommend these precautions if you live in an area black bears are known to inhabit.

At Home

- **Please don't feed the bears**. Ranger Smith in the Yogi Bear cartoons of the 1950s had it right after all. You may love watching black bears devour a pile of nuts, doughnuts, or candy bars from the safety of your living room window, but this practice does more harm than good to the bears. Bears who become conditioned to come into people's yards to find food soon lose their fear of humans, leading them to wander through neighborhoods and towns looking for treats. They may even approach random people expecting handouts, a scary proposition for unsuspecting residents who do not encounter bears regularly. Inevitably, this leads to the bears' untimely euthanasia. Please don't play a role in this.

- **Store trash receptacles inside a sturdy building** that you can lock securely: a garage or reinforced shed, rather than an open shelter or carport. If you don't have a garage, purchase a single trash can enclosure like the ones available from Bearsaver.com or Bearicuda.com, and lock your trash can inside daily. Most of these are made from steel rather than plastic, creating a formidable barrier against bears.

- **Use bear-proof trash cans** with locks that bears can't open. If your trash collection company requires you to use a specific rolling receptacle that they supply, you can still protect your trash with aftermarket latching devices that prevent bears from gaining access.

- **Secure your recycling**. Empty cans and bottles can retain the aromas of the food they held even after washing. Store these in a bear-resistant container or in your locked garage. If your community requires you to bring your recyclables to a recycling center, do this at least once a week to defeat the accumulating food smells.

- **Wait until morning on trash pick-up day** to put your trash on the curb. If you are using bear-proof trash cans, unlock the locks as close to pick-up time as possible.

- **Clean your grill**. After every use, scrub the grate and the inside of your grill to eliminate lingering smells and leftover food particles. Wipe down attached shelves, trays, and deck floorboards to get rid of juice or sauce drippings. If you use a smoker, don't leave it unattended for long periods while using it, and lock it up inside your garage when it's not in use.
- **Feed birds in winter only**. Bird seed, nuts, hummingbird nectar, suet, and grape jelly are all irresistible to bears. If bears live in your area, feed birds only from late fall to early spring, when bears are in hibernation. Birds most need the support of feeders in winter, when natural food sources disappear just as they require more protein and fat to maintain their body temperature.

    You can still attract birds year-round by planting native flowers, shrubs with berries, and trees that produce berries or nuts. These may also attract bears, but they do not provide the calorie boost that bears seek when they plunder human trash cans and buildings. If you intend to plant shrubs with berries or nuts, place them well away from your house and garage to keep bears from exploring them for food as well.
- **If you do feed birds, use bear-proof feeders**. Seed feeders made of steel instead of mesh or plastic can withstand some rough handling by a black bear. Fully enclosed with metal, they usually have windows so you can see when the seed has run low, and fewer openings and perches than conventional feeders. A thick, steel loop at the top keeps a bear from pulling the feeder off its hanging wire.
- **Bear-proof your bird-feeding station**. Use heavy-duty steel cord to hang your bird feeder, making it very difficult for a bear to chew through it. Consider suspending your feeders at the top of a high pole where bears cannot reach them—some available feeder poles have a crank that allows you to lower the station to your height when it's time to refill the feeders. Some handy bird lovers rig up a pulley system that places the feeders high up in a tree or on a tall pole, lowering them easily to fill them, clean them, or

take them down in spring. Black bears are agile climbers, but they can't climb a pole with nothing for their claws to grip.

- **Keep pet and livestock food indoors.** Do not leave bowls of dog or cat food outside on your porch or in your yard. Feed your pets indoors and lock up bags of pet food or livestock feed indoors as well, as bears will break into (or even push over) a small shed. Bears will come right up onto your porch to feed on these treats, increasing the risk for you and your family.

- **Keep appliances indoors.** No matter what you do to lock and secure it, a refrigerator on a porch or patio in bear country means free lunch to a black bear. Bears can smell what's inside a closed refrigerator and will do just about anything to get at it, including pushing the fridge over and rolling it until it opens. You'll have a mess on your hands unless you keep this appliance inside your home or locked garage. (It goes without saying that leaving a portable cooler outside—even an empty one—essentially hands it over to the bears.)

- **Keep packaged food indoors.** A screened porch is not an indoor setting; bears can tear through screens with little effort. Store canned, plastic-wrapped, and foil-wrapped foods in your house, and keep the doors and windows closed and locked at night. Bears' claws can rip into cans, pull caps off of bottles, and tear any food wrapping. Packaged foods within bears' reach present another hazard: bears do not always recognize that wraps are not part of the food, so they may eat plastic, paper, and foil, which can make them very ill.

- **Clean your vehicle.** Food wrappers and cups left in your car still smell like food, even though the food is long gone. Bears can smell gum, mints, granola bars, scented lotion or hand sanitizer, lip balm, soft drinks, coffee, and any other edible or product you keep in your car. Get these leavings into your bear-proof trash can as soon as you get home, and keep toiletries in a purse or briefcase, so you take them with you when you leave the vehicle. Even with this precaution, your car can continue to smell like the burger you ate yesterday. Use a scented air freshener (as long as it smells like pine or new car, and not like food) or an upholstery freshener to eliminate lingering food smells. Above all, lock your car, especially if you park it outside overnight.

- **Lock your dog door at night.** Dogs often need the ability to come in and go out as they please, but in rural or heavily wooded areas, this opening can provide access for all kinds of wild animals, including young bears. Secure the door each evening and overnight to keep from finding a bear in your kitchen first thing in the morning.
- **If you compost, use a bin.** Compost bins are designed to contain the odors associated with decaying material, making them an excellent option in bear country. Follow the rules of responsible composting and keep meat, dairy products, and large amounts of fruit out of your bin.

In the Neighborhood

- **Talk to your neighbors.** You can do everything exactly right to keep bears from visiting your property, but if your neighbor doesn't play by the rules, bears may still frequent your yard. Tell your neighbors—especially people who have just moved in—about the bear activity in your area and share the best ways to bear-proof their home and yard. If you are part of a homeowners' association (HOA), bring this up at a meeting and encourage the HOA to establish a formal policy to discourage bear activity.
- **Protect your dog.** If you have a dog, it's easy to believe that they will help protect you in a confrontation with a black bear, but recent research proves otherwise. Canadian professors Stephen Herrero and Hank Hristienko examined black bear incidents throughout North America that took place between 2010 and 2014, and found that in 53 percent of the attacks, dogs were involved. "In the vast majority of cases, it seemed as through the

dog(s) had been running loose at the time of the attack and drew the bear to their owners," Herrero told the *Calgary Herald*.

Female bears with cubs may be more easily provoked into a confrontation when a dog is present, this research determined. Most often, this happens when the dog is not on a leash, whether the dog and its owner are out walking or are even in their own backyard.

A recent incident in Butler County, Pennsylvania, provides a dramatic illustration of this. Lee Ann Galante had just let her eight-pound Pomeranian out in her own backyard when the dog began to bark, and Galante spotted three bear cubs in her neighbor's tree. She had no time to react before the mother bear appeared and came after Galante, knocking her down. "It was just so fast," Galante told the ABC News affiliate in Pittsburgh from her hospital bed a few days later. "She was so aggressive. She was all over me. She had my dog. She bit my arm. She was on top of me." At one point, the bear took Galante's head in its jaws, breaking bones in her face.

The dog, Smokie, continued to bark and distracted the bear long enough for Galante to scoop him up and get inside. She slammed and locked the door, narrowly escaping what had seemed like inevitable death.

The bear left her with injuries to her neck and arm and the likelihood of future surgery to repair her facial bones, but she is expected to make a full recovery. Smokie got away with a minor scratch. The Pennsylvania Game Commission located the mother bear and found her still acting aggressively, so they captured her and euthanized her. Her cubs were believed to be mature enough to manage on their own.

Why would the presence of a small dog put an adult bear on the defensive? "When species share space, competition often develops," writes Masterson. Bears see dogs as a threat, no matter what size they are or whether the dog barks or behaves aggressively toward the bear. "And person plus dog may be more alarming to a bear than either one by itself and could trigger a defensive reaction," she adds.

To maximize your and your dog's safety if you live in bear country, always keep your dog on a short (six feet), non-retractable leash when you are out for a walk. Keep your eyes and ears open so you can observe and listen to your surroundings. Letting your dog roam free, no matter how well trained you believe it to be, can invite trouble if there are bears nearby.

- **Make noise when hiking.** Talk, sing, call out, or clap your hands at regular intervals. Some outdoor outfitters sell bear bells that jingle gently as you walk, but bears have only average hearing, so these generally are not loud enough to let the animals know you're on your way. Once a bear hears you coming, chances are good that he or she will move away from the trail and leave you alone.
- **Assume that bears are nearby.** You don't see them, but don't assume there are no bears in heavily populated neighborhoods or on local trails. Keep making noise (and ignore the people who give you the evil eye for being noisy) and keep your eyes open for bears in the area.
- **Watch out for surprises.** If your neighborhood has streams, shrubs full of berries, fields of cow parsnip, lots of oaks with acorns, or areas of dense vegetation, keep your eyes open for bears. As they can't always hear you, you may startle a bear by arriving quietly.
- **Do not approach bears.** No matter how often you see them calmly eating berries off of local shrubs, bears are not tame, and they are not zoo animals. You are in their natural habitat, so steer clear of them as much as possible.
- **Skip the selfie.** Don't turn your back on a bear to take a selfie, and don't try to get closer for a better photo. You don't want the last photo you take of yourself to be of a foolish person with an obviously agitated bear in the background.
- **Carry bear spray.** Pepper spray is one good defense against a charging bear. Nontoxic and with no permanent effect, it triggers

"temporary incapacitating discomfort" in the bear, which can halt an attack and give you the opportunity to get away. If a bear charges you, aim the aerosol directly in the bear's face. This is not a bear repellent—spraying it on yourself (as you would an insect repellent) will not keep bears away.

- **If you encounter a bear, do this.** As every bear will react differently, there is no set protocol that will result in a sure-fire escape. Try to detour around the bear if possible.
- **Do not run!** Talk calmly so that the bear can tell you're human and therefore not a threat or prey. Make yourself look larger and move to higher ground if you can. Do not climb a tree; black bears are good climbers and may follow you. Drop something (not food) to distract the bear. Keep your pack on for protection in case of an attack. If a bear attacks and you have pepper spray, use it!

The mountain lion (also known as cougar, puma, panther, catamount, or mountain cat) is much more numerous in its range than most homeowners believe.
© MAZIKAB, ISTOCK

Chapter 2

SECRETIVE NEIGHBORS
MOUNTAIN LIONS

Saying that you've seen a mountain lion (*Puma concolor*) in New England these days has all the credibility of saying you've seen an ivory-billed woodpecker in Arkansas or a Sasquatch . . . well, anywhere. In 2011, however, a fairly impressive cross section of Northeastern residents swore with absolutely certainty that they had spotted a mountain lion in a number of places near Greenwich, Connecticut—crossing a backyard, skirting the edge of a schoolyard, or moving along the edge of a highway with all the stealth this large, wild cat is known to muster.

The stories seemed impossible to New England wildlife officials, as the region's very last mountain lion died in Maine in 1938. The federal government attempted to codify the extinction of the Northeastern lion subspecies—known specifically as the Eastern cougar (*Puma concolor cougar*)—back in 1973, but one diligent wildlife researcher, Bruce Wright, convinced authorities to categorize it as "rare and endangered" rather than nonexistent. He provided a long list of anecdotal sightings by New England residents over the previous several decades as "proof" of the cat's existence—but this was long before everyone had a camera in their pocket, so he had no photos or film to accompany any of these reports. Eventually, the US Fish and Wildlife Service (USFWS) took the final step, declaring the Eastern cougar extinct in 2011 and finally removing it from the endangered species list in 2018.

Call it what you will—puma, cougar, panther, painter, catamount, or one of forty other names for the animal worldwide—a mountain lion comeback still seemed possible to some of its fans over those ensuing decades. Professional tracker John McCarter came upon evidence of big cat scat in the Quabbin Reservoir in New Salem, Massachusetts, in 1997, much larger than the far more common bobcat could produce. When he searched the area for additional evidence, he found a "cat cache," body parts of a recently killed beaver buried to keep other predators from stealing it. DNA tests of scat, hair, and blood at the site proved that they were produced by a mountain lion, though what race or subspecies this cat might have been remain open questions to this day.

Fourteen years passed before any other credible evidence came to light. In March 2011, just days after the USFWS announced the cougar's extinction, professional forester Steve Ward photographed fresh tracks in the snow across a frozen-over cove in the Quabbin Reservoir. A team of experts from throughout the region and across the country examined his high-quality photos, and they agreed that only a mountain lion could have made the tracks. While members of the general public took this to mean that a new population of mountain lions had emerged in New England right under the noses of experts reporting their permanent demise, the experts did not agree. Instead, they cited mountain lions' ability to wander long distances, even across many states, to find an environment that provides what they need to survive. The closest populations of mountain lions to Connecticut lived in the Dakotas, and a small number of Florida panthers—another endangered species—struggled to keep their population intact in the Southern states' forests.

Residents continued to report sightings, however, including one who spotted a mountain lion on the grounds of a private school on June 5, 2011, in Greenwich, Connecticut.

At about 1:00 a.m. on June 11, 2011, an animal darted out in front of a sport utility vehicle driving on the Wilbur Cross Parkway north of exit 55 in Milford, Connecticut, about forty miles east of Greenwich. The driver, taken by surprise, struck the animal and killed it. The driver called the police, who collected the 140-pound carcass of what was obviously a mountain lion, a creature that had not been sighted for certain in New England in more than seventy years.

State wildlife officials began to piece together the unlikely circumstances that placed this wild cat in Connecticut after all this time. Their first theory postulated that the lion had been kept illegally as a pet by an unknown, secretive resident, but a criminal investigation turned up no evidence of this. Meanwhile, scientists and mountain lion specialists from New England to the Rocky Mountains pooled their knowledge and data, enlisting the help of the US Department of Agriculture's Forest Service Wildlife Genetics Laboratory in Missoula, Montana. The results were nothing short of amazing: this mountain lion had come from South Dakota, leaving its DNA in Minnesota and Wisconsin in scat, hair, and blood in 2009 and 2010 while scientists there tracked it. The young lion, judged to be about five years old, had made a journey of more than 1,500 miles—only to have its life end in an instant on a New England highway.

The lengthy trek was "more than twice as far as the longest dispersal pattern ever recorded for a mountain lion," wrote Peter Applebome for the *New York Times* News Service on July 27, 2011, when the results were announced.

This trail of irrefutable genetic code gave Daniel C. Esty, commissioner of the Connecticut Department of Energy and Environmental Protection, the opportunity to underscore his department's ongoing message: "This is the first evidence of a mountain lion making its way to Connecticut from Western states, and there is still no evidence indicating that there is a native population of mountain lions in Connecticut."

This, however, has not stopped the debate over the existence of mountain lions in New England. In the days that followed the South Dakota cat's death, two people reported lion sightings: one in a homeowner's yard, and the other near the Merritt Parkway. With no photos, video, or other proof, however, these claims became easy for the Department of Environmental Protection (DEP) to dismiss out of hand. Nevertheless, the sightings persisted, until the *Record-Journal* of Meriden, Connecticut, reported on August 28, 2011, that a whopping forty-seven sightings had been reported since the first one on June 5. Most of these turned out to be a deer, coyote, or bobcat—a much smaller cat with a spotted coat and

Mountain lions need forested land and plenty of deer to survive. When they lose these necessities, they may come in closer to humans to prey on livestock.
@ MAZIKAB, ISTOCK

stubby tail (see chapter 3)—but some residents stuck to their stories even when they could not produce proof of what they had seen.

Just in case one of these accounts turned out to be accurate, however, the DEP issued instructions to residents of areas near wildlife refuges about precautions they should take. They warned residents to keep their young children and pets close, and to go inside at dawn and dusk when mountain lions are most active.

Where Did the Lions Go?

Mountain lions across North America depend on two things to keep their numbers strong: forests in which they can hunt with their legendary stealth; and plenty of deer, their number one choice of prey. The big cats had ample amounts of both in the eastern half of the continent until the mid-1700s, when Europeans began to arrive in large numbers to claim the land and build towns and cities. By the mid-1800s, the heavily forested White, Green, Berkshire, Adirondack, Allegheny, Appalachian, and Blue Ridge Mountains stood denuded, their hardwoods turned into lumber to construct homes and municipalities. "In 1850, the slopes would be anything but green," wrote Paul Bierman in his 2008 paper on clear-cutting and erosion in New England, published in the journal *Vignettes*. "Most would be barren, stripped of their trees, and trampled by grazing sheep. Deep gullies and shallow landslides were common. Some of this eroded material ended up in river channels choked with sediment but much of it went no farther [than] the bottom of the hill from which it was eroded."

In creating living spaces for so many immigrants, the construction boom left nowhere for mountain lions and other woodland creatures to live.

A fascinating timeline provided by the Mountain Lion Foundation allows us to track precisely what happened to America's mountain lions. With their forests gone and their deer sharply limited, some lions struggled on, skulking in wooded areas alongside farms and helping themselves to livestock—calves, lambs, and chickens—at every opportunity. This attempt at survival only angered the cats' new human neighbors, however, and as early as 1684, the new colony of Connecticut offered a generous bounty of twenty shillings (about $290 in 2025) for the dead carcass of a catamount. Other colonies followed suit, and after the Revolutionary War, states and territories across the continent joined in. By 1845, hunters had wiped out mountain lions in Ohio altogether, the first state to lose its whole population of them. Lions in Indiana, Illinois, Massachusetts, Iowa, and Vermont fell to bounty hunters over the next thirty years, until there were no more of them in these states either.

Even in the Western states, where smaller human populations stretched over many miles of open land, mountain lions were classified as predators and hunters were rewarded for bringing them down. The Utah Territorial Legislature went so far as to classify the lion as an "obnoxious animal" in 1888, following this with a bounty of its own a few years later. Killing of the last mountain lion in each state expanded south and west, until Wisconsin, Kansas, South Dakota, and Missouri declared them extirpated in the early 1900s. Even the US Fish and Wildlife Service participated in the wholesale slaughter, working with the state of Utah in 1917 to protect livestock by hunting and shooting the lions.

Passage of the ironically named Animal Damage Control Act in 1931 knocked the last nail into the coffins of a wide range of predatory mammals, providing the Department of Agriculture the permission it needed to protect livestock and cultivated land to the exclusion of everything else. The law called for "the destruction of mountain lions, wolves, coyotes, bobcats, prairie dogs, gophers, ground squirrels, jackrabbits, and other animals injurious to agriculture, horticulture, forestry, husbandry, game, or domestic animals, or that carried disease." States embraced this new law and increased the size of the bounty per mountain lion, offering fifty to one hundred dollars for every animal killed and delivered to authorities. From 1947 to 1969 in Arizona alone, hunters received rewards for killing more than five thousand cougars.

Then in the late 1950s, wildlife management organizations began to look around and notice that no more mountain lions roamed their forests. As the slow realization finally dawned on the nation's people that killing wild animals, stripping landscapes, and polluting air and waterways could harm the planet irrevocably, the state of Florida became the first to pass legislation to protect its endangered panthers. Other states shook themselves awake as well: Utah, Oregon, Washington, Montana, California, and Colorado all dropped their bounty programs when they found that the number of animals turned in had taken a sharp dip. People had stopped killing mountain lions because there were no more mountain lions to kill.

Faced with this unintended but entirely predictable consequence of the bounty system, the United States reversed its position on animals in general in 1966. Passage of the Endangered Species Preservation Act gave mountain lions and many other animals their first, best chance at survival since Europeans had arrived in North America. When the Eastern cougar subspecies landed on the list in 1973, Western states still classified mountain lions as game animals—but they ended open season on them, restricting when they could be hunted to a few weeks a year.

Despite a massive reforestation effort in the Eastern states, Eastern cougars never returned to the now-shady mountain landscapes from

Maine to Georgia. The Western states have fared better, thanks in part to legislation that protects mountain lions in several states, including a landmark law in California that reclassified lions as a "specially protected mammal," ending hunting of them for good. The Mountain Lion Foundation worked to see passage of this law in 1990 and has continued its efforts to protect the cougars throughout their Western territory. With reduced hunting and similar protections taking hold in the Midwest, mountain lions have been seen again in Nebraska, Missouri, Wisconsin, Illinois, and South Dakota—the state from which the lone lion made its trek across the Eastern states to a Connecticut highway in 2011.

Living in Lion Country

Sitting on the porch of the Fall River Visitor Center in Estes Park, Colorado, in August 2018, I interviewed Don Hunter, science director of the Rocky Mountain Cat Conservancy, about a mountain lion's involvement in the 1997 death of a ten-year-old boy, Mark Miedema, on Rocky Mountain National Park's North Inlet Trail. (The story of this death appears in my book *Death in Rocky Mountain National Park*.) Hunter noted that the boy's encounter was one of less than twenty attacks by mountain lions in North America in the last hundred years.

"It's not normal for lions to eat people," Hunter said. "They don't think of us as prey. When the natural prey of big cats becomes diminished, they tend to seek out domestic livestock or sometimes small pets, not humans."

The child, running down the trail ahead of his parents, had done nothing unusual to attract the lion's attention. A park spokesperson suggested to the Associated Press that "Mark came along at the wrong time when the lion was in a hunting mood."

The lion emerged in a flash and darted across the forest, grabbing Mark as it crossed the path and dragging him into the woods before his parents had any opportunity to react. They ran after the lion and fought for their son, frightening the animal enough that it dropped the boy and darted off. Mark's wounds were severe, but they were not what killed the boy. In his terror, he choked on his own vomit.

The park tracked down the lion and killed it, but the incident made many a Colorado parent's blood run cold at the thought that their own child could become prey to one of the big cats in their area. They needn't have worried about another attack, said Hunter. Lions eat a variety of prey, with small mammals and deer at the top of their list. They fear humans and do their best to remain unseen, moving quietly and remaining secretive in the shelter of forests.

"There's a lion within half a mile of us now," Hunter said, and my husband and I could not help but turn in our seats and look around, just in case. "It's the most widely distributed mammal in the world, because it's in both North and South America. They've lived with people for a very long time, and they don't like people. Lions here are no different than anywhere else."

Hunter came to Colorado after Mark Miedema's death, when the park sought his expertise in large wild cats—he had spent much of his career researching snow leopards in Asia. "Our study showed the mountain lion population is healthy, especially in Estes Valley—there's lots and lots of food," he said. "We cataloged the first lion in 2003, and found it spends as much time outside of the park as inside the park. . . . We have tracked our lions right to the edges of town; they're around people all the time. So if the lions saw people as prey, they would be eating someone every day."

The chances of actually seeing a mountain lion on or near property frequented by humans are very slim, though the presence of livestock or pets can increase the odds—but only if your ranch or farm has forested sections where lions can conceal themselves easily. "The potential for being killed or injured by a mountain lion is quite low compared to many other natural hazards," the National Park Service tells us. "There is a far greater risk, for example, of being killed in an automobile accident with a deer than of being attacked by a mountain lion."

Authorities including the National Park Service, the Mountain Lion Foundation, and my own books offer these recommendations for staying safe in mountain lion country:

- **Don't hike alone.** As attractive as a solitary hike may seem, it's a dangerous proposition for a whole host of reasons, including the possibility of an animal attack. If you do hike in a group, keep children in sight at all times, and discourage them from running ahead of you.
- **Wear bright-colored clothing.** Make it obvious that you are not the lion's usual prey. Lions do not normally see humans as food, so wearing colors that it does not associate with deer or other woodland animals will differentiate you enough to warn the lion to leave you alone.
- **Look for signs and other postings.** Parks and trail maintenance organizations often put up warning signs when mountain lions have been seen in a particular area. Check carefully before heading out and ask rangers at visitor centers about which trails to avoid.
- **Don't jog or ride a mountain bike through forests at dawn and dusk.** In these crepuscular times, mountain lions are particularly active, so your swift movement could attract interest from this predator.
- **Keep pets close.** A six-foot maximum leash will keep your dog from wandering off and accidentally rousing a mountain lion. If you know you're hiking in lion country, you may be better off leaving your pets at home.
- **If you see a mountain lion:**
  - **Don't panic.** Stand straight up and face the lion, and don't run away; wild animals will pursue a runner automatically. If you must move, back up slowly until you are out of sight of the mountain lion. If you have small children with you, pick them up if you can, to keep them from running away.
  - **Do not move toward the lion.** This will seem like a confrontation to the lion, increasing the possibility that it will become defensive. If you stand still, the lion will most likely just walk away.

- **Stay tall.** Scientists have found that mountain lions don't recognize a standing human as prey—so if you stand straight, they are unlikely to approach. If you crouch down, however, you look a lot more like a four-legged animal, and their perspective will change.
- **Intimidate the lion.** If the lion starts to approach you, do everything you can to look large. Raise your arms above your head, wave your arms, flap your jacket open, and yell at the animal. If this does not work, throw stones, branches, or whatever you have in your backpack in its general direction—not to hit it, but to hit the ground in front of it. This shows the lion that you can defend yourself, which may be enough to scare it off. Don't try to hit the lion; an injured lion will not make your situation better.
- **Protect yourself.** If the lion still approaches you, you have the right to defend yourself. Throw things to hit it—a water bottle, book, or whatever you can reach without bending down.
- **If the lion attacks, do not play dead. Fight it as hard as you can.** Use whatever you have, from your hiking poles to your fists. A charging lion will go for your neck or head, so use whatever you have in your hands to keep it from doing this. Your fellow hikers can help as well. A full-grown mountain lion weighs 130 to 150 pounds, or roughly as much as the average human. While the lion's teeth and claws give it a significant advantage, it is not so large that it can't be wrestled off of you, especially if you're part of a group.

Keeping Your Livestock Safe

- **Keep your pets indoors at night.** Lions are most active from dusk to dawn, so leaving your pets outside unattended could attract a mountain lion to your backyard. If you need to let your dog out at night, keep it leashed and stay close to it until it is ready to come back inside.

- **Keep your livestock in a shed or barn at night.** If you have a small farm with a few animals, it may be possible to round them up and keep them sheltered from dusk to dawn. For larger farms or ranches, keeping all livestock inside may not be possible—so the Mountain Lion Foundation suggests these steps to protect your investment:
  - **Clean up any blood on your property.** Blood from birth, sickness, or injured animals can attract mountain lions and other predators.
  - **Place vulnerable animals in a covered pen.** Newborn sheep, goats, llamas, and calves may need additional protection from lion attacks. A number of predator aversion pens are available commercially, including directly from the Mountain Lion Foundation at https://mountainlion.org/stay-safe/#!enclosures-and-pens.
  - **Install a very tall fence.** Mountain lions can jump straight up 15 feet, so a higher fence may be required.
  - **Employ guard dogs.** Some dogs are trained to ward off predators including mountain lions, scaring them away before they can attack.
  - **Use motion sensor alarms.** You can frighten off a wide range of predators by using motion-activated alarms with bright lights, loud noises, and even sprinklers.
  - **Talk to your neighbors.** Even if you're taking every possible precaution, a careless neighbor who does none of these things can still ruin it for the whole neighborhood. Talk to your neighbors about deploying the same methods, making your entire area inhospitable to lions and other potential predators. If you work together, it can be possible to eliminate predator attacks for the long term. The Mountain Lion Foundation offers literature and other tips for getting neighbors together at https://mountainlion.org/wp-content/uploads/2021/06/Door_Notice.pdf.

Bobcats have made themselves quite at home in residential areas, where they may show up in backyards or in the middle of a city. © CHRIS BOSWELL, ISTOCK

Chapter 3

FIERCELY INDEPENDENT
BOBCATS

Deep in the heart of Texas's Dallas-Fort Worth Metroplex, an area that long since traded its oak forests and backland prairies for a glittering cityscape of chrome, glass, concrete, and mowed lawns, a population of unlikely residents has made its home between the state highways, the intersecting railroad tracks, and the Trinity River. Bobcats (*Lynx rufus*) find this urban landscape oddly appealing, moving stealthily along narrow green belts in residential neighborhoods and through the riparian area along the river.

Why are they in the Metroplex? Urban settings attract rodents, the primary food source for bobcats, so these wild cats have become an integral part of Dallas-Fort Worth's residential ecosystem. Bobcats hunt rats and mice, rabbits, hares, songbirds, ducks, snakes, squirrels, opossums, an occasional deer fawn, and other small animals, keeping populations of some nuisance critters from expanding beyond control. This useful purpose to human residents has made bobcats a fairly welcome animal in this north-central Texas city, and their nocturnal habits keep them comfortably out of sight.

With its thick, heavily spotted coat, striped legs, and bobbed tail, the bobcat is fairly easy to tell apart from common domestic cats. Bobcats tend to weigh in at fifteen to thirty pounds and may be as much as forty inches long from ears to stubby tail, making them somewhat larger than healthy house cats. That difference, however, does not stop some human neighbors from assuming that these solitary animals have the same interest in being around humans as domestic cats do. When a few of these cats began to congregate in one area, some Dallas-Fort Worth residents called on animal control, asking for the removal of their "problem" bobcats.

Richard Heilbrun wasn't ready to take them at their word. As state wildlife diversity program director for the Texas Parks and Wildlife Department, he objects to the overwide definition of "problem" that many human residents apply to animals.

"I ask, 'What is the animal doing that you feel is a problem?'" he told me in a Zoom interview. "Usually, they tell me it's out in daylight, and it's making some of their neighbors nervous. That doesn't rise to the level of a problem."

Most bobcats have no interest in people, he said, so they won't approach a person and will keep to themselves when humans are around. "That's why when people see a bobcat, it's noteworthy. They get excited that this amazing animal lives near a creek in their city."

The encounter changes, however, when people try to influence a bobcat's behavior, especially if they start feeding the animals. "Problem animals are in the eye of the beholder," Heilbrun said. "Left to themselves, the bobcats may not be a problem—but if you're feeding them, then that may become a problem."

That's what happened in a neighborhood in one of the most tightly developed parts of the Metroplex, he noted. "People were feeding bobcats in other areas, and word got around that this particular neighborhood had a lot of bobcats. So to get rid of the bobcats in one neighborhood, they trapped them and brought them over to this neighborhood. Now you have habituated bobcats brought into an even worse situation."

Bobcats expect to eat what they have freshly killed on their own, so it takes some effort to get them to eat something offered by humans. Once they associate people with food, they may become bolder, or they may congregate in one backyard—an unnatural situation for them. If they begin competing for food in a small area, the wild cat might pass up its usual rodent dinner in favor of someone's guinea pig or chihuahua.

And that, said Heilbrun, makes it a problem. "If they take a pet, if they scare a kid, if a neighbor is scared to walk to her car because some-

In some cities, it's not uncommon to find a bobcat surveying a residential area as it waits for a rodent to reveal itself. © WELCOMIA, ISTOCK

one has five bobcats in their yard, now we have a situation where we need to know more."

Heilbrun became part of a team of researchers who examined the Dallas-Fort Worth bobcat population to understand more about their numbers and what they mean. In a study published in 2019 in the journal *Urban Ecosystems*, lead researcher Julie Young of Utah State University and four other wildlife specialists, including Heilbrun, created a system to determine how many bobcats actually lived in an area of about seventy-eight square kilometers (roughly thirty square miles). They focused on densely developed parts of Fort Worth, Arlington, Hurst, Bedford, Euless, and Grand Prairie, where residents saw bobcats often.

The team mounted pairs of trail cameras opposite one another alongside roads or trails near suitable bobcat habitat. For six to seven consecutive weeks at each camera station, they photographed bobcats and studied the photos to learn the distinctive spot patterns that identified individuals. Some of these cats had been fitted with GPS collars to monitor their movements before the researchers undertook the study, so this data helped with sorting out the various bobcats as well.

To get the most accurate calculation of the number of bobcats in the monitored area, Young and her team used a method known as spatially explicit capture-recapture (SECR). This model estimates the density of a specific species while taking the location of its capture (in this case, its photograph) into consideration. This method has been used in many studies to improve the precision of population estimates of tigers, leopards, and lions in studies in Africa and Asia, but it had never been used to estimate the density of carnivores in an urban environment. The team found SECR effective in determining that an estimated 1.28 bobcats per square kilometer lived in the survey area, for a total of about forty-three bobcats across the grid.

"Our study revealed a high density of bobcats in an urban landscape despite most assumptions that bobcats require large areas of habitat and are sensitive to fragmentation," the study concluded. "A robust population of bobcats in the heart of a dense metropolitan area, like DFW, provides optimistic possibilities for the potential of bobcats and other carnivores to thrive in urban landscapes with minimal conflict."

In other words, bobcats have found ways to coexist with all of the development around them, moving through this congested landscape without the benefit of wildlife corridors in much of the area.

This understanding provided just one answer to the many questions about bobcats in Dallas-Fort Worth, however. What were the bobcats eating? Why were some of them congregating in groups, a completely new behavior for these loner cats? Why were some showing themselves in daylight? In short, how was city life changing the lifestyle of the bobcat population?

In their years of monitoring these bobcats, Heilbrun and the research team collected many scat samples and analyzed them to see what the cats ate. Other information came to them through lay observations of bobcats with prey detailed on iNaturalist, and a few deceased bobcats yielded the opportunity to analyze the contents of their digestive systems. The results revealed no surprises: the remains of rats and mice were found most frequently, and birds made up the second most frequent meals. Citizen science observers reported many bobcats with squirrels in their jaws, though scat analysis turned up rabbit and the occasional opossum, raccoon, or nutria.

"In scat analysis, we found zero anthropogenic sources," said Heilbrun. "In analysis of stomach contents, we found zero human food sources."

So, despite their modern surroundings, bobcats in Dallas-Fort Worth had continued to make their living the way they always had, feeding on rodents and birds they killed themselves and staying away from trash cans, pet food, or any other human discards.

Wild cats in other neighborhoods and other Texas cities, however, had changed their habits. Some had become accustomed to accepting food from humans.

"There's no healthy, normal reason for a bobcat to interact with humans," Heilbrun concluded. "When that happens, that's a direct indication of people trying to keep them."

In a neighborhood in San Antonio, for example, residents reported two bobcats that were out in broad daylight, thinking this might be an indication that they were rabid. Nothing was amiss with the cats, as it turned out, but a landowner, thinking the bobcats were having trouble finding food, had started feeding them ground meat. This lured the cats out of their daylight habitat for an easy meal.

"If a bobcat doesn't have to earn a living, why would it?" said Heilbrun. The landowner received a visit from the state wildlife agency, and the bobcats, finding no more easy pickings on his property, went back to hunting their own prey.

They're Everywhere

Despite disappearing forests in much of the Eastern United States and dense areas of human development, bobcats continue to live in some part of nearly every state—with the notable exception of Delaware, according to the International Society for Endangered Cats. While they prefer wooded areas, they also find a comfortable living in arid scrubland landscapes, coastal swamps, and Western deserts, avoiding the agricultural lands of the Midwest where they do not find the cover needed to protect themselves from wind, weather, and hunters. Their adaptation to urban areas allows them to maintain a much larger population than their close American cousin, the far-rarer Canada lynx.

This is not to say that bobcats have not faced their share of challenges. They vanished from some states in the mid-1800s as settlers and industries felled entire forests for fuel, lumber, and charcoal, leaving few areas for bobcats and other wildlife to live and raise their young. Farming took hold of much of the cleared land from the Eastern states through the Great Plains, altering the environment so much that bobcat populations virtually disappeared. Bobcats still attempting to hold their own in their accustomed territory became prey for fur trappers as bobcat pelts turned into a very valuable commodity, and for sport hunters looking to add to their trophies. By 1850, the state of Ohio declared the bobcat extirpated. Other states recorded significant declines in bobcat sightings as well, driving trappers north and west to continue trading their skins.

With the marked decline in bobcats, however, came a corresponding increase in rodent pests, especially in agricultural areas. This led farmers to set traps bated with rat poison—and some of the few remaining bobcats caught and ate poisoned rats before the rats themselves could die from the poison. When the cats ate these rats, they ingested the poison, which killed many of them. It seemed the human world had conspired specifically to eliminate bobcats, though the real culprit was the lack of understanding of the consequences of human action.

None of these circumstances put the brakes on bobcat hunting or trapping. In the 1900s, states with large numbers of ranches, dairy farms, and chicken farms established a bounty for every bobcat killed by a hunter or farmer, making it profitable on several levels for farmers to shoot them on sight. A surge in the price of bobcat pelts in the 1970s intensified the fervor to trap the cats by any means necessary—including the use of spring-loaded steel jaw traps, which cause animals so much pain and suffering that they have since been banned in some states.

These issues so depleted the bobcat population across the country that some states took action to protect them. Connecticut reclassified the bobcat as a protected furbearer in 1972, ending their hunting and trapping there for the foreseeable future. Ohio followed suit in 1974, and when the state removed the bobcat from the endangered list in 2014 as the cats returned to some parts of Ohio, it kept in place the ban on hunting and trapping them. California's Department of Fish and Wildlife established a bobcat hunting ban on January 1, 2020, while the department studies the current population and gains a better understanding of its health and needs.

By contrast, Indiana passed a controversial law in March 2024 to allow open season on bobcats beginning July 1, 2025, just nineteen years after the animal was delisted as a state endangered species. "No studies, estimates, or educated guesses on the bobcat population were presented during either the Senate committee hearing or the House committee hearing," Fox59 News in Indianapolis reported.

Despite the different approaches, one thing all states' wildlife experts agree on is that bobcats are not a threat to people, even if they make an appearance in a resident's backyard. They do recommend some precautions to keep pets, livestock, and the bobcats themselves safe if they do visit your property.

- **Do not feed animals**. Bobcats have no interest in bird food, but they may be attracted to the birds, mice, rats, groundhogs, squirrels, and other small animals that feed on seed, nuts, or suet. Bobcats also may be attracted to your birdbath, especially in a desert or semi-arid environment. You can reduce the possibility of a bobcat hunting under your feeders by regularly sweeping or raking up the seed and shells that fall to the ground, so small animals don't congregate there.
- **Feed your pets indoors**. This may not be possible with every dog or cat, but leaving pet food out on your porch is a signal that animals feed here often. If you must feed your pets outside, do so during daylight hours and clean up thoroughly afterward, so leftover morsels don't attract small, wild animals.
- **Don't leave your pets alone outside**. A small dog or cat is an easy target for a bobcat, especially if there is no human nearby. Having a fenced backyard is not enough protection, as bobcats can leap effortlessly to the top of a six-foot fence.

- **Keep your pets and livestock indoors from dusk to dawn.** This is especially important for small pets like rabbits, gerbils, hamsters, guinea pigs, chickens and other poultry, and so on. During the day, an outside pen will protect them if it has a secure lid, and if all seams of the enclosure are wired together securely.
- **If they must be outside, keep them in an enclosure with a woven wire or electrified fence.** The California Department of Fish and Wildlife suggests a fence with two electrified wires at twelve inches and eighteen inches above the ground, as well as on fence posts and any other structures. One mild shock will be enough to discourage a bobcat.
- **Repair your fences.** Fix holes right away, as these are easy entrances for bobcats and other animals you may not want in your yard.
- **Keep shrubs and grass trimmed.** While piles of twigs and brush are wonderful cover for birds, bobcats use these brushy areas for cover while hiding or resting. A well-tended yard will deny them that respite.
- **Use scare devices to startle the bobcat away.** Motion-sensor lights and noisemakers can help on a temporary basis, until the bobcat learns that they are not actually dangerous. Spraying them with a garden hose or motion-activated sprinklers can be effective as well.
- **If a bobcat shows up in daylight**, go indoors and watch the cat through a window for signs of rabies: agitated behavior, aggression, drooling, foaming at the mouth, and loss of muscle control. A healthy bobcat will simply go about its business and eventually wander off, but if you see issues that may signal disease, call your community's animal control organization immediately (usually by dialing 911).
- **If you encounter a bobcat on a trail**, it will most likely run away from you immediately. In the unusual event that it stays and looks at you or begins to approach, make a lot of noise to scare it off.

Back up slowly; don't turn and run, as this will initiate the cat's instinctive behavior of running after prey. If this still doesn't work and the bobcat continues to advance, spray it with water (squeeze your water bottle at it)—and in the highly unlikely event that it attacks, fight back as hard as you can. Bobcats are larger than house cats but not by much, so do your best to get your hands around it and throw it away from you.

Tom turkeys can be very impressive—and intimidating—when they display during breeding season. © NIC MINETOR

Chapter 4

BACK WITH A VENGEANCE
WILD TURKEYS

It started with nine.

No one seems to know the name of the woman who released nine "pet" wild turkeys (*Meleagris gallopavo*) on the grounds of the South Beach Psychiatric Center on the New York City borough of Staten Island in 1999. If she truly chose them as pets, it's anyone's guess why this woman kept wild turkeys under her care instead of acquiring the bright-white domestic turkeys raised on farms—the kind that are fed to virtually everyone in the country on Thanksgiving.

The hospital, adjacent to Staten Island University Hospital (SIUH), Ocean Breeze, is on the East Shore, about two blocks from New York's Lower Bay. The turkeys may or may not have had a connection to the psychiatric hospital at some point—sources differ on exactly what that link may have been—but the grassy grounds provided enough seed, grubs, and shoots for the foraging birds to scratch out a living there.

Suddenly freed after a lifetime of domestication, the turkeys did what all hardy birds do: they multiplied. Turkeys are particularly adept at this, laying clutches of ten to fifteen eggs per breeding season and raising most of those to adulthood, the four-foot-tall birds using their five-foot wingspan to intimidate cats, rats, foxes, and other carnivores to keep them from poaching their young.

The birds' population soon outgrew its home on the hospital grounds, spreading outward through eastern Staten Island. As they had begun their lives under human supervision, the original Ocean Breeze nine had no fear of people—so they passed on this relaxed attitude to their broods. The offspring, in turn, launched their own lives away from their parents as young animals do, seeking their own patch of fertile ground to sustain themselves and to start their families. A quarter of a century later, Staten Island finds itself in the midst of the turkeys' own version of urban sprawl. Neighborhoods are loaded with flocks of these enormous brown birds, dominating public parks, scrapping with pets, nesting in residential yards, blocking auto traffic, roosting in trees, defecating wherever they choose, and generally annoying the stuffing out of the human population.

Recently, a new chapter in the lives of New York City's turkeys has begun. Staten Island may not be big enough to hold all the birds it has produced; in 2024, turkeys began showing up in Jersey City, New Jersey, just north of Staten Island. And on May 7, 2024, the Manhattan Bird Alert posted to X that a wild turkey had paid a visit to a neighborhood near Park Avenue and Forty-Ninth Street, taking refuge in a tree as night fell in the nation's most populated city.

What, if anything, must be done? Can people learn to live with these large, domineering neighbors, or will local and state wildlife officials do something to reduce the turkey population?

A Checkered Past in New York

Wild turkeys lived in New York state for millennia before its statehood, eventually serving as a food source for the Indigenous population hundreds of years before the earliest European settlers arrived. With the influx of immigrants in the eighteenth and early nineteenth centuries, forest-dwelling turkeys lost habitat to clear-cutting, making them easy prey for hunters seeking to feed their families. Soon, however, more affluent settlers arrived with different sensibilities about hunting, seeing its value as sport while farms and city markets provided all the meat they needed to eat. Turkeys fell in large numbers to hunters who killed indiscriminately for their own pleasure. By the mid-1840s, wild turkeys were nearly extirpated from New York State. The last wild turkey in Massachusetts fell to a hunter's bullet on Mount Tom in 1851, and historic accounts suggest that it may have been the last one in all of New

England. Soon, native turkeys were gone altogether from the Eastern half of the United States.

A century later, however, with regulations in place that limited the number of turkeys a hunter could take in a season, wild turkeys began to return of their own accord. They crossed through forested areas along the Pennsylvania–New York state line, beginning to repopulate western New York and soon appearing in grassy fields on farm properties and in wooded patches between farms. Encouraged by this natural emigration, the state's Department of Environmental Conservation began a formal reintroduction program in 1959, bringing small populations of turkeys into other areas throughout New York State. Today, the state reports that some 180,000 turkeys live within its borders, "including some that have made their home right here in New York City," the Wildlife NYC website reports.

Wild turkeys feeding on lawns in residential areas have become a point of contention in Staten Island, New York, and in other cities across the country. © NIC MINETOR

Exactly how many of these turkeys live in Staten Island appears to be unknown—no agency reports have ever attempted to count them—but to many residents, the only answer is "too many." Turkeys had become so comfortable with their human neighbors that they blocked roads, ignored traffic stopped to let them pass, and confronted homeowners who tried to shoo them off of their front lawns.

So back in 2016, the New York State Department of Environmental Conservation (NYSDEC) joined with the US Department of Agriculture (USDA) and SIUH to begin a program of capture and relocation of the birds. As reported by Staten Island media, the "milestone deal" focused on the turkeys that lived on the hospital campus, where their "overwhelming presence on the hospital grounds pose[d] considerable safety risks to our patients and visitors," said SIUH executive director Donna Proske to SILive.com at the time. "Our concerns stem from aggressive birds near our facility entrances, traffic stoppage from large flocks interfering with ambulances and patient drop-offs, and unsanitary eliminations on the sidewalks."

The turkeys would be transported to a farm upstate—an actual farm, not the euphemism so many children hear when they lose a beloved pet—an arrangement hailed by local and state government officials as "a humane conclusion" to the ongoing nuisance. "When government bureaucracy is responsive to community needs, we can achieve productive results," said Congressman Daniel Donovan, whose constituency included Staten Island and Brooklyn, in an interview with SILive.com.

But two years later, the And-Hof Animals Sanctuary for Farm Animals + Permaculture had taken in more than 150 turkeys from Staten Island, and the twelve-acre refuge had reached the limit of what it could hold. The farm also provided (and continues to provide) a home for goats, cows, sheep, chickens, pigs, and other animals rescued from substandard or abusive conditions. Even beyond capacity issues, however, a larger problem had become clear: the 150 relocated turkeys were a small fraction of what remained on Staten Island. "We are concerned because they will continue to breed and the problem won't stop," said Kurt Andernach of And-Hof Animals to SILive.com. "It's self-perpetuating."

Back with a Vengeance: Wild Turkeys

NYSDEC and USDA required And-Hof to create a species-specific area for the turkeys, so the sanctuary had graded land, built a pond, and fenced a large enclosure, a costly endeavor for the small farm. "From the beginning, there was never any money offered," said Andernach, and requests for funding from NYSDEC had been refused. Bringing in more turkeys would mean a larger enclosure and more materials and labor, none of which the sanctuary could afford without assistance.

So the relocation operation came to an end, and Staten Island's turkeys continued their normal rate of propagation. On the *Staten Island Advance*'s "From the Scene" podcast in 2022, Erik Bascombe interviewed the *Advance*'s public interest and advocacy reporter Kristen Dalton. "The state has now taken a stance where they say that they don't believe that relocation is the best solution to the turkey problem," Dalton told Bascombe. "But they are open to conversations to figure out what to do."

Dalton reported that there has never been an official turkey count, but during the 2020 to 2022 pandemic, people noticed the turkeys out and about more than usual—and the population definitely increased while fewer people crossed the city on a daily basis. Turkeys began to appear in neighborhoods where they had not been seen before. Residents complained more often about property damage, especially to their cars. "One car looked like it had been hit up by bullets," said Dalton. "There were scratches and peck marks all over the car. Then they're going in bushes and shrubbery in front of houses and damaging that."

City councilmembers and other officials continued to look for a solution, but the most recent advice is that residents should learn to live with the flocks of big birds as they would any other wildlife on the island. "Unless you get every single turkey that's on Staten Island, they're going to keep having babies," Dalton said. "And there's just no way to know if we got all of them, especially ... because they're all over Staten Island now."

This is not all bad, of course. Birds like turkeys serve an important purpose that benefits people: they keep down the population of insects, especially ticks, which carry a variety of diseases that are highly injurious to humans. They can even be considered entertaining, strutting their way down city sidewalks like portly shoppers, or gathering in groups for a lively gobble-fest. Occasionally, the birds become territorial and pick fights with one another, battling it out in the middle of a city street. (In one video, a resident can be heard yelling, "Turkey fight! Turkey fight!" Turkeys may provide Staten Island with as much fun and excitement as a WWE episode.)

Now that they are a quarter-century into the turkey situation, state and local officials have "thrown in the towel" on trying to find a solution that removes the birds from the borough's streets and yards. In the most recent reports, the state DEC has "deferred responsibility" for the Staten Island turkeys to NYC Parks, the city-wide supervisor of public spaces. When I contacted NYC Parks, press officer Kelsey Jean-Baptiste responded by email: "NYC Parks does not manage or track wild turkeys." The office does, however, use reports made to the Wildlife NYC website (https://www.nyc.gov/site/wildlifenyc/things-to-do/report-a-sighting.page) to "monitor interactions between wildlife and the public."

She concluded: "Wild turkeys are one of the many native species that call our city home. If you spot a wild turkey, don't be intimidated—just keep your distance and observe respectfully."

Is There Another Way?

One question appears repeatedly in group chats, Reddit comments, and other social media frequented by Staten Islanders: Why not trap and kill them, and provide the meat to shelters and community pantries to help alleviate hunger in our cities?

No one I tried to contact would discuss this, including the National Wild Turkey Federation and the turkey specialist at NYSDEC, among others. A simple search on news stories about Staten Island's turkeys revealed the answer, however: despite the number of complaints about the birds, some Staten Islanders love them. So when a blue T-shirted crew from the USDA showed up on August 12, 2013, on the South Beach Psychiatric Center campus and began luring the turkeys into large black nets, witnesses reacted with nothing less than horror.

"They were picking the turkeys up by their necks and feet, and throwing them into plastic crates," one witness told the *Staten Island Advance* on August 13. The workers piled the crates four-high in the back of two pickup trucks "and then they just let them sit," she said.

USDA spokesperson Carol Bannerman portrayed the scene a bit more clinically. Calling the event a "direct assistance" operation requested by the psychiatric center, she said, "USDA biologists and specialists are removing free-ranging wild and hybrid turkeys from the campus." The turkeys "are herded into temporary corrals made of netting, hand-captured, and placed in poultry crates. The birds are then transported to a state-approved processing facility. The resulting meat will be stored frozen until testing confirms its suitability for donation for human consumption."

The drastic measure was necessary, she continued, because of the "concern for excessive feces accumulation on handrails and walkways," and the obvious hazards to health and safety that this presented.

Despite all of the complaints that preceded this action, however, Staten Island residents wanted none of this. They held a protest on a street corner near the psychiatric center, waving signs that proclaimed, "God Loves All Creatures!" and "Educate, Don't Decimate," while politicians addressed the crowd, calling the cull "ridiculous" and demanding to know why there had been no public notice before the roundup. Scathing editorials criticized NYSDEC, which had provided the permit for the slaughter after refusing to relocate the turkeys. Many of these birds, NYSDEC said, had interbred with turkeys living on farms, creating hybrid strains that could taint the natural bloodline of purely wild turkeys. If they could not be relocated and they could not continue on the South Beach campus, culling the flock became the only option.

"We may lament that it came to this, but consider that every November, the same thing happens, only on a much larger scale," the editorial board of the *Staten Island Advance* wrote on August 15. "We can only hope that these birds can ultimately be used to feed people who can use a good meal."

This was by no means a sure thing. As part of its Wildlife Damage Management Technical Series, the USDA provides extensive information about wild turkeys and the hazards they may cause to crops, property, and human health and safety. While it steps back from suggesting that wild turkeys are inedible, it enumerates the diseases and parasites that these birds can acquire. "Some common infectious diseases includer avian pox, Lymphoproliferative neoplasms (transmissible tumors), infectious sinusitis, and histomoniasis (blackhead disease)," the document's writer, Mississippi State University Professor Emeritus James E. Miller, tells us. "Although these and other infectious diseases are sometimes found in wild turkeys, none are known to pose a human health threat."

If you don't find that particularly comforting, just wait. "Wild turkeys are also hosts for a variety of internal parasites, including protozoans, trematodes, cestodes, acanthicephalans, nematodes, and arthropods," he says. "External parasites found on turkeys include ticks, mites, lice, and louse flies. Some of these can be transmitted to people." His recommendation: "Wear appropriate protective gear, such as rubber gloves, when handling live or harvested turkeys."

The National Wild Turkey Federation (NWTF) looks to assuage fears around the harvesting of wild turkey meat as well. "Diseases that affect wild turkeys are not a threat to people or domestic animals," writes James Earl Kennamer, PhD, on the NWTF website. In fact, NWTF has offered events at which chefs teach people how to prepare a harvested wild turkey for cooking, and how to cook the turkey (with some very tasty-sounding recipes).

The fifty-two birds that met their fate in August 2013 did provide hearty meat to people in need, but the outrage that this operation created among the island's residents put an end to further discussion of such a step. A month after the slaughter, twenty-eight birds that remained on the psychiatric center's grounds were relocated to a new home at Catskill Animal Sanctuary, with NYSDEC granting them a reprieve (what the papers called a "pardon") in response to the public backlash.

Today, Staten Islanders do their best to live in harmony with the giant birds, while the birds make very few concessions to their human neighbors, continuing to strut their stuff on streets, in yards, and in parks wherever and whenever they choose.

Boston and Beyond

Staten Island is not alone in the challenges it faces with resident turkeys. After a very successful program in 1972 to relocate just thirty-seven wild turkeys from western New York State to the Berkshires and other areas in western Massachusetts, where turkeys had been extirpated more than one hundred years earlier, turkeys continued to proliferate, moving east through the state under their own power and with the help of wildlife professionals. Eventually, they found their way past the I-495 beltway and into the city of Boston.

Like New York, Massachusetts now has a spring and fall turkey hunting season, which keeps the wild population in check from year to year. The ones that chose city life, however, essentially outsmarted hunters by collecting in flocks in residential areas of Brookline, Cambridge, Falmouth, and Boston—places where hunters are not permitted to fire a gun. Here turkeys have made a comfortable home among humans, even though some people are not very comfortable around them.

"We never reintroduced turkeys into Eastern Massachusetts proper," MassWildlife biologist David Scarpitti told Boston.com in 2023. "They just got there on their own. I don't think anybody really expected turkeys to become so common in these hyper-developed landscapes."

Scarpitti offered quite a bit of advice to Boston-area residents for dealing with their turkey neighbors. First, he noted that turkeys maintain an actual pecking order in their groups, choosing a dominant male turkey as the leader of their group based on the bird's physical size and strength. This bird will lead the others to places to forage for food. In mating season, with hormones raging, the lead bird may behave aggressively toward anything that approaches it: dogs, cats, humans, even cars.

"Because they're not afraid of humans, they see another animal within their territory and they're going to act aggressively towards it to establish their role in that pecking order," Scarpitti said.

This is the behavior that famously led Benjamin Franklin to call the turkey a "Bird of Courage, [that] would not hesitate to attack a Grenadier of the British Guards who should presume to invade his Farm Yard with a red Coat on." (Popular legend says that Franklin proposed the turkey as the official bird of the United States, but this founding father's suggestion that the turkey would have been a better choice came in January 1784, a year and a half after the bald eagle had already been selected and made part of the national seal.)

Turkey antics can be intimidating, especially when the bird pulls itself up to its full four-foot height and flaps its wings or pecks in a human's direction. The trick to overcoming this, Scarpitti said, is to not play the weaker role. Most people do not encounter a large, seemingly angry bird on a daily basis, so the sight of one can scare them into turning around and taking another route to wherever they're going. This gives the bird a win over the person, which will only make the turkey more assertive in the next situation.

"Stand your ground," said Scarpitti, even if that seems like a precarious choice. The bird may attempt to circle a person to gain an advantage, with the intent of attacking from behind. Keep the bird in front of you, wave a broom or open and close an umbrella in its face, or raise your arms high over your head—or even flap your arms like wings. If the bird perceives a person as a threat, he will retreat and take his gaggle elsewhere.

Even this confrontative approach may not work every time, as Tess Bundy of Brookline discovered in April 2017. The Associated Press reported that a large tom turkey "launched itself at her and her infant daughter, backing down only after Bundy whacked it several times with a shovel." Bundy herself told the AP, "Every year they're worse. I really do think that they're a menace to the town."

Indeed, in Brookline, several turkeys had become so aggressive that police had to shoot them.

It's the same on roadways, when turkeys stand in the middle of the street and block traffic; a car is just another big being to attempt to dominate. "You just have to keep moving," said Scarpitti. "The turkey is not going to lay down under your tires and let itself be killed." Moving ahead slowly but consistently will convince the turkey that it must get out of the way or give up its life to the big metal animal.

With turkeys making a spectacular comeback from a time when an estimated fifteen thousand remained throughout the United States, other states have had to cope with turkeys becoming urban dwellers. Iowa, Oregon, and Montana have banned the feeding of wild turkeys in an effort to reduce their numbers in residential areas, and communities across the country including Casper and Buffalo, Wyoming; Bowling Green, Ohio; and many others have taken up the issue in their city councils. In Buffalo, New York, feeding wild turkeys is now a misdemeanor with a $750 fine.

"We have no way of halting this," said Terry Asay, head of the Buffalo Planning Department, to the *Buffalo Bulletin* as the city considered the ordinance. "People are out there feeding them, and they've become such a nuisance. The ordinance will put a little bite behind the bark. We can issue a citation if people are caught feeding them, in hopes of discouraging it."

Some communities have attempted another solution: watching turkey nests carefully for clutches of eggs, and "addling" the eggs—shaking them, puncturing the shell, or rubbing oil on the outside of the egg—to make the embryo inside unviable. Shaking or puncturing the egg destroys the embryo, while coating the egg with corn oil prevents the exchange of oxygen while the embryo incubates, essentially suffocating it. The egg is placed back in the nest, so the adult turkey continues to attempt to incubate it, preventing her from laying a new clutch that year.

Wildlife professionals across the country have used this method for nearly sixty years to discourage nesting of birds in areas where they destroy property or create a public safety concern. Ring-billed gulls, double-crested cormorants, Canada geese, herring gulls, and mute swans have all been targets for addling, as these birds can gather in very large flocks that crowd out less aggressive species, and cause damage to buildings and property or to agricultural fields. Only professionals with permits from the official management agency for a particular species can carry out an egg addling operation, as there are strict rules about when, where, and how addling can be done.

How to Live with Wild Turkeys

Turkeys that have become habituated to coexisting with humans are very unlikely to retreat into more appropriate habitat. If the city they inhabit becomes less hospitable to their presence, however, they will be less likely to make nuisances of themselves for local residents. Follow these simple rules to keep turkeys from dominating your life and neighborhood:

- **Don't feed the turkeys.** The most basic rule is also the most effective. Turkeys don't require people to feed them to survive, but like all animals, they will take the path of least resistance to a full belly.

Do not offer them birdseed, corn, grain, or any other kind of food that will attract them to your yard. If you do feed songbirds and other perching birds, use foods that do not leave shells and other scraps on the ground—feed with suet, sunflower hearts (no shells), peanut pickouts (no shells), and nyjer seed. Avoid seed blends with millet and milo, as most small birds reject these, so they pile up on the ground and attract other critters like turkeys (as well as mice and rats). Do not use cracked corn or any blend with corn in it.

- **Keep turkeys from nesting in your yard.** When you see turkeys on your property, chase them away. Making noise (banging a metal pan with a spoon, for example), clapping your hands, and spraying them with a hose are all good ways to drive them out of your yard. Keep an eye out for their return and harass them again until they stop showing up. Consistency is key to making your property less attractive to them.

- **Keep your pets indoors or on a leash.** Dogs and cats can't help but challenge turkeys when they approach. It may seem like a barking dog will drive away a turkey, but the birds may respond by charging or pecking your animals. If a dog encounters a whole flock of turkeys, nothing good will come of it for your dog.

- **Do not approach turkeys.** Despite whatever nuisance they may present, turkeys are a lot of fun to watch ... from a respectful distance. Whether they're crossing the road in front of you or they're on the other side of an open field, your safest choice is to stand back and away from them. This goes double for breeding season, roughly from April to mid-June.

- **If turkeys approach you, scare them off.** Make big gestures, wave your arms and yell at them. A turkey may try to establish its dominance by charging at you or pecking in your direction. Show them that you are bigger, stronger, and louder than they are to encourage them to keep their distance.

- **If all else fails, get professional assistance.** In some states, only a licensed nuisance wildlife control professional can touch, trap, or remove a wild animal from your property. Contact your state's environmental conservation or fish and wildlife office to see what services may be available in your area.

From the edges of wooded areas to lawns, parks, and even cemeteries, deer are reclaiming areas that they once grazed freely. © NIC MINETOR

Chapter 5

SO. MANY. DEER.

In 2015, the small city of Tega Cay, South Carolina—a 2.5-square-mile suburb of Charlotte, North Carolina, with a population approaching fifteen thousand—began to notice a change in its resident population of white-tailed deer (*Odocoileus virginianus*). Coyotes had kept the deer herd at a manageable level for many years, but recently, the gentle animals had begun to reproduce at startling rates.

"Years ago, a prior city council made the decision to begin removing coyotes from the city due to negative interactions between coyotes and pets," said Tega Cay city manager Charles Funderburk in a November 2024 email conversation. Hunting is not allowed in the city, so with the last natural predator gone, the suburban neighborhoods became particularly safe spaces for deer and their offspring. Deer tend to thrive when they share the landscape with humans, because people create ideal areas for them. Our beds of abundant flowering plants and shrubs, our "forever wild" forests just off of our backyards, and the edges between agricultural fields and woody bushes all invite browsing deer to partake of a gorgeous four-season buffet.

Some residents who enjoyed hosting these quiet visitors went so far as to place deer feeders in their backyards, making these stations favorite gathering places for animals in the mild Carolina evenings. Soon Tega Cay saw more and more deer within its residential areas—deer who ate the buds off of their native shrubs and trees, devoured their tulips and other tasty perennials, and chewed tree seedlings down to the ground.

This abundant food supply makes for healthy bucks and does, which in turn make plenty of healthy fawns. In December 2015, the South Carolina Department of Natural Resources (SCDNR) conducted its first spotlight survey of the deer population and came back with a manageable number of individuals: just 156 (or sixty-seven deer per square mile), noting, "This number would be high for a wildland situation; it is not unusual in a residential situation." While SCDNR only had the staff to conduct the survey on one night—normally at least two nights would be required—the city council at the time accepted the results and took no action to control deer numbers.

Over the next few years, residents could not ignore the herd's steady growth. Not only did the deer devour gardens, but they also ran through traffic, forcing motorists to stop on a dime to avoid running into them. A deer strike can cause as much damage to a car as to the deer, and in some cases, the driver can sustain injuries as well.

By 2020, the city's wildlife management program stopped removing coyotes from the peninsula, allowing natural predation to resume. This appeared to be too little too late, however, as the deer continued to reproduce and raise young at an increasing rate. "City Council enacted a law to prohibit the intentional feeding of wildlife," said Funderburk, eliminating deer feeders from the equation, but this was not enough to reduce the population.

In January 2022, complaints about free-roaming deer devouring gardens, ruining hedges, and defecating on lawns made it clear that the Tega Cay City Council needed to take more aggressive action. This began with a mission to gather better information about the size of the herd. SCDNR conducted two spotlight surveys in March and September and discovered that the population had grown to about seven hundred deer in March—and by September, that number had expanded to 850, or 301 deer per square mile.

It was time to look carefully at the range of solutions available. The council inquired about relocating some or all of the herd to a park or wildlife refuge, but they found that both the Department of Natural Resources and People for the Ethical Treatment of Animals (PETA) strongly recommended against this. "They said of all the options in reducing the herd, this was in their words unethical and inhumane," said Funderburk to the *Rock Hill Herald*. The risk to the deer was unacceptable, these organizations said, as deer are not cattle—they are not accustomed to being transported in trucks or on trains. The animals could be injured or even killed in transport.

Logically, hunting seemed like a more practical option. The Tega Cay City Council put out a survey to its residents with one question: "Would you be in favor of the City utilizing sharpshooters as a method to managing the deer population?" Residents responded, but the results didn't help. They were almost perfectly split on the issue: Half (1,106) said yes, and half (1,010) said no. Media reports about this survey asked residents why they voted against the culling, and most of them noted that firing guns in residential areas is simply a bad idea—not because deer would be killed, but out of concern for their neighbors and children. "There are a lot of citizens out at night who are not capable of making the best decisions," said Michele Coburn to the *Rock Hill Herald*. "There's nobody that I trust to shoot moving objects at night, with [children] asleep 15 feet away."

Mary Ickert, another local resident, agreed. "Let's face it: there's a reason there are rules in place not to allow hunting in a residential area," she said.

When another survey in September 2023 found that steady growth had continued over the ensuing year, with the deer population rising to 1,028 deer (or a staggering 349 per square mile), the city council made the decision to cull some deer from the herd. An extensive budgeting, permitting, and ordinance-amending process finally wrapped up in January 2024, when US Department of Agriculture (USDA) sharpshooters joined with local police and sheriff's department officers at the Tega Cay Golf Course to begin hunting for deer on six designated days in January and February. The $100,000-plus effort had a mandate to take as many as 160 deer—80 per three-day hunt. The meat harvested from these deer would go to local food banks.

This should have been like the proverbial shooting of fish in a barrel—a clear field of vision with few obstacles, plenty of prey, and fairly close quarters. By the end of the six days of hunting parties, however, the hunters had only taken 36 of the intended 160 deer.

What went wrong?

Funderburk offered this explanation: "As this was the first time in the city's history we have allowed for high powered rifles to be intentionally fired, we put a number of safety controls on the USDA staff that was contracted to do the culling," he said. "Those controls really limited where and when they could harvest deer, which ultimately led to fewer deer being culled."

Thirty-six deer removed from the herd would not lead to a noticeable reduction in their numbers, so Funderburk and his council began to look closely at what he calls a "hybrid approach": more culling, but with the addition of a sterilization program for a selection of two hundred does.

"Our herd currently has a 24 percent reproduction rate annually," said Alicia Fornato with the Tega Cay Wildlife Conservation Society to the *Queen City News* in August 2024. "So, there is no way to kill our way to success with numbers like that."

The added tactic of preventing the does from reproducing would make a difference over the long term, allowing natural attrition and a slowed birth rate to take the numbers down humanely.

To accomplish this task, the city hired White Buffalo, a nonprofit organization that specializes in helping cities and towns control the population growth of native species.

White Buffalo explored two different methods of sterilizing: birth control medication and surgery. As of this writing, some birth control drugs for deer are in the testing phase, to be sure that they are safe for humans that might consume a medicated deer. Other "fertility control agents" are available to control deer reproduction, according to White Buffalo's website, but the deer must get more than one dose of these after a specific interval, making them very costly to administer (as much as $1,500 per doe in medication and personnel) and difficult to use in a wild animal population.

The most dependable and cost-effective method, then, is surgical sterilization: using tranquilizer darts to put the doe to sleep, and then quickly removing her ovaries in a fifteen- to thirty-minute procedure. This "one-and-done" approach renders the deer unable to conceive in every case. White Buffalo's staff takes great care to keep the doe comfortable until she can be released back into the wild, using a sleeve over her eyes and snout to calm her. Once the ovaries have been removed in a minimally invasive procedure, technicians place numbered tags in the doe's ears that can be seen from a distance, helping to keep her from becoming a target when culling begins later in the season. Some of the does also receive radio collars to help track their movements through Tega Cay.

In a whirlwind twelve days in October and November 2024, White Buffalo captured, sterilized, tagged, and released the promised two hundred does. While this method's effect on the size of the herd will not be felt for several years, sterilization has generally been more acceptable to the city's residents than culling the herd, said Funderburk. "Unfortunately, the public still seems pretty split 50/50 on the culling," he said. "We can't reduce the deer population by just sterilizing (at least not anytime soon) and culling to reduce isn't a long-term fix either. In conversations with White Buffalo, we think the hybrid approach that we are taking this year, along with natural attrition, will get us to a more manageable number sooner rather than later."

WHEN PREDATORS ROAMED THE CONTINENT
Before Europeans moved into North America, Indigenous people shared the continent with predators that fed on hooved animals like bison, elk, mule deer, white-tailed deer, and pronghorns. Gray wolves roamed in packs, working together to take down young, weak, or lame animals lagging behind the herd. Red wolves did the same in the Southeastern United States, keeping in check the animals that feed on shrubs and grasses—which in turn allowed this vegetation to thrive and guaranteed a substantial food supply for deer from year to year. Lush riparian areas grew up along the lengths of rivers and creeks, providing food for everything from ducks to rabbits to beavers. Balance reigned in natural areas from Maine to California.

Mountain lions—known as cougars or pumas in much of the country—played a critical role in this balance, making their stealthy advances on their prey and easily separating the ailing or slow members from their herds. Mountain lions are especially adept at this, stalking some prey for hours or even days until the young deer's endurance for the battle simply runs out. The people with whom they shared the forests, fields, and plains saw the big, secretive cats only occasionally, but they recognized the critical role they played in the animal kingdom, ensuring that only the strongest deer, elk, and others lived long enough to reproduce.

Pursuit by predators sent deer and elk to roam far and wide to keep ahead of wolves and lions, a practice that prevented them from overgrazing in any one open area. Instead of lingering in open fields, they browsed on the edges of covered areas like thickets and deep forest where they could shelter themselves quickly from potential attack. Trees, shrubs, and grasses thrived alongside the grazing animals, renewing the natural crops each spring and providing a never-ending food supply.

When Europeans arrived in North America, they began building farms and ranches, introducing different kinds of animals into the vast, open landscape. Christopher Columbus brought the first cattle to the Americas in 1493, but several centuries passed before large herds of domestic cows grazed the land. Likewise, archaeologists believe that the Spanish and Portuguese settlers in the 1500s established the first farms with domestic chickens in residence, but raising chickens as an industry developed far later, alongside transportation methods like trains and canals. Suddenly, a cattle or chicken farmer could transport meat and eggs to markets more than a few miles from their own homes, so big farms and ranches became profitable alternatives to family farms for the first time.

With big livestock farms came the need to protect calves and chickens from natural predators. Suddenly wolves and cougars were no longer seen as important players in maintaining a healthy countryside—instead, they became farmers' worst enemies. "European settlers feared and hated these animals," notes the Ohio Canid Center website. "The livestock they brought with them on settlements was left to roam freely—and were found by wolves to be easy meals. Fear and misunderstanding of wolves quickly led to hatred."

By the latter half of the nineteenth century, wolf and cougar hunting became common practices among farmers and ranchers. Not only did farmers hunt down and shoot these animals, but they came up with another way of eliminating them: poisoning animal carcasses left out in fields to lure these predators in. This killed the wolves and cougars while preserving their hides intact, with no bullet holes, to sell to the lucrative pelt trade.

The US Forest Service led the charge to rid the agricultural world of wolves in the early 1900s—and in just a few decades, gray wolves were completely eradicated across the contiguous United States. Mountain lions disappeared from most of North America as well. (See chapter 2 for more about lions.) With these predators gone, attention turned to coyotes, which now raided chicken coops and pastures with greater abandon since their main competitors were out of the picture. Coyotes became a target for extermination as well, though they have managed to thrive even at the objections of livestock owners. Today, thanks to programs like Project Coyote that promote coexistence with wild carnivores, coyotes are not in danger of going the way of wolves in North America. (More on coyotes in chapter 7.)

Mule deer in Western neighborhoods take the same liberties as white-tailed deer, grazing in the open on mowed lawns. @ NIC MINETOR

Deer as a Primary Target

Logic suggests that with no mountain lions or wolves to prey on deer and diminished numbers of coyotes in human-populated areas, the deer population must have exploded. Indeed, millions of deer roamed the continent in comparative safety from their natural predators—but they became prey to a much more numerous and deadly adversary.

Open season for hunters also extended to deer and other hooved animals throughout most of the nineteenth century. Deer provided food for hungry families, while their hides served as a convenient source of fabric for clothing. Deer hides became a source of income for trappers and hunters as well, who sold them into the European trade for such goods. Westward expansion made deer hunting even more important to homesteading settlers, who lived almost exclusively off of what they could grow, trap, or hunt in their immediate area. Restrictions had just begun to emerge in Eastern states on the time of year or the number of deer people could take—but in the latter half of the 1800s, the newly formed Western states had no hunting laws on the books.

Things got even more dire for deer after the Civil War, when many new kinds of hunting aids became available to civilians. Traps, pitfalls, snares, and repeating rifles made hunting more consistently successful, and the use of dogs and night lighting increased. With all of these tools at their disposal, hunters wiped out more than 95 percent of the nation's deer by the 1890s, leaving fewer than five hundred thousand white-tailed deer across the continent—in fact, some estimates say barely more than two hundred thousand—where an estimated twenty-four to sixty-two million had roamed before European settlement.

"In fact, deer in the United States were limited to inaccessible areas in northern Maine, pockets of the Adirondacks, southeastern Massachusetts, the Ozarks, and undeveloped regions of the Gulf Coast and the Mid-Atlantic states," notes David G. Hewitt in an article published in the *International Journal of Environmental Studies* in 2016.

Realizing the dire situation, departments of natural resources in many states moved to rein in the wanton killing, establishing limits on how many deer could be taken per hunter, and during which seasons hunting could take place. Across the country, states began to establish game agencies that led the efforts to reestablish local deer populations, working with landowners to bring deer from remote locations onto their wooded land. Game preserves emerged—notably, the Pisgah National Game Preserve in western North Carolina played a critical role in bringing deer back to several Southeastern states, using land donated in 1915 by the estate of George Washington Vanderbilt, heir to a portion of the family's railroad fortune. Blooming Grove Hunting and Fishing Club in Pennsylvania, the Adirondack League Club in northern New York, and Vanderbilt's Biltmore Estate—each with a membership of enthusiastic hunters—all joined the effort to raise white-tailed deer and repopulate the species across their regions. Many other clubs and organizations joined the effort. Today an estimated thirty million deer live in the nation's forests and other wild areas, and well-regulated hunting keeps the population at a healthy level.

What these conservation-minded organizations did not foresee, however, was what would come with restoration of deer numbers across the continent: the mounting competition from deer for grazing and browsing land.

Deer began to propagate at a rate that soon rose from a pleasant surprise to a worry, then a problem, and right on to a nuisance. The growth of their herds took time, as most deer had not moved from the forests into residential areas where hunting was prohibited until the 1990s, so hunting seasons still kept their numbers in check. As urban areas begat suburbs and suburbs began to sprawl, however, human development began to butt up against and replace accustomed deer habitat. Wooded areas gave way to mowed lawns and small clusters of trees. Natural understory disappeared, and carefully planned hedgerows offered deer lots of edges—lines of woodland separated by open fields and mowed yards for eye-level browsing. Deer adapted readily and eagerly to this new, accessible habitat and all of its tasty snacks. Without the restoration of the natural predators that had historically kept their numbers in check, deer could bring all the young they could produce into this new, safe world.

Meanwhile, climate change has created more hospitable territory for deer across the country. Herds that used to move into more densely vegetated areas for the winter now find all the food and shelter they need wherever they lived from spring through fall. They have become year-round residents in neighborhoods where plantings offer them all the buds, berries, bark, and bird seed they need to get them through the colder months—and in some of these areas, residents invite them to linger at feeders designed especially for deer. Now when hunting season rolls around in late fall or early winter, deer may remain in areas filled with human homes and playing children, so firing a hunting rifle is illegal.

So now we have significant populations of deer living in very close proximity to humans, where they readily wander into front yards, cross the roads in front of oncoming traffic, and otherwise endanger their own health and safety.

There's another issue emerging in recent years that makes deer populations particularly hazardous for their human neighbors: the spread of illnesses borne by ticks. The list of these continues to grow at a fairly alarming rate: Lyme disease, babesiosis, anaplasmosis, ehrlichiosis, hard tick relapsing fever, and Powassan encephalitis are all diseases that can be spread to humans through a bite from the right kind of tick. Not all of these ticks are found in every region of North America, but they are carried by deer and other hooved animals, and they have spread rampantly in some states. Check with your local department of natural resources for the latest information on which tick-borne illnesses may be present in your area.

How to Live with Deer

Most of the objections to deer in human residential areas have to do with garden preservation rather than direct encounters. Keeping deer away from gardens full of tasty shoots, leaves, and bark can be tricky, but it can be done.

- **Annoy the deer enough that they leave.** Some people start the process of discouraging deer by spreading commercial deer repellent materials. These usually have an odor that deer don't care for, like mint, garlic, thyme oil, rosemary, rotten eggs, or "multiple smell and taste deterrents," as one of these products defines its contents. Whether or not these work depends in part on how appealing your plants are to deer, and whether they are willing to brave the stench to reach your delicious leaves and berries. Most deer repellents are not harmful to plants or animals, but you'll want to check the package carefully before spreading them to be sure they are safe for pets or around human-edible crops.

- **Motion-activated sprinklers** can have some effect on deer and other animals, hitting the deer (or the neighbor's dog or cat) with a blast of cold water whenever animals enter the yard. What makes this solution effective is its consistency, turning on every time it detects motion or heat, and saving homeowners hours of vigilance with a hose or squirt gun in hand, waiting for the offending animals to show up. Deer that get tired of being wet every time they approach your garden may simply stop coming. Some deer are cleverer than others, though, and will simply stay beyond the reach of the sprinklers once they determine what the splash pattern is. Keep moving the sprinklers around your yard to confuse the smarter deer.

- **Scare the deer away.** A short-term solution may be to simply frighten the deer enough that they do not approach your property again for some period of time. Some homeowners have used homemade scarecrows to unnerve deer, while others try mylar streamers, balloons, or other shiny objects that are outside of a deer's frame of reference. Ultrasonic repellants emit a sound at a high frequency that only animals can hear, driving away deer (and many other animals) for short periods. These are most effective especially if they are motion-activated, so every time the deer approaches, it hears this annoying burst of noise. Deer will eventually come to understand that these devices will not actually hurt them, however, so they may return to your tasty trees and plants before much time has elapsed.

- **Install a fence.** The most direct and effective way to discourage deer from eating your plants and trees is to prevent them from entering your yard at all. A deer-proof fence must be at least eight feet high, notes the Nova Scotia Department of Natural Resources

(DNR) and Renewables in Canada, where deer are especially plentiful. Others recommend an even higher fence, reaching ten feet or more. The most deer-resistant fences are made with woven wire, but this can be very expensive, the DNR says—so only "very high value or high-risk areas" like open cropland or airport runways can truly justify the cost. "For small gardens, one might improvise with a discarded ground fisherman's net or snow fence," Nova Scotia suggests. The DNR also notes that some gardeners or farmers have had success with fences that slant outwards, making them harder for a deer to surmount in a single leap.

- **Consider an electric fence.** Less expensive than you might think, electric fencing provides a sure-fire deterrent to browsing deer. It only takes one or two encounters with electric shock before deer become conditioned to stay away from the barrier. The deer's natural inclination is to squeeze under a barrier rather than leap over it, so that jarring encounter will most likely happen shortly after you install the fence. Anecdotes from farmers include recommendations of hanging some kind of bait on the fence to lure deer in for that one lesson: Suspend an apple on a string from one of the wires or smear peanut butter on a sheet of aluminum foil and fold it over a wire at nose-height for an adult deer. Deer will come right over for the perceived treat, feel the shock, and often back off from the fence for good.

- **Plant things deer don't like to eat.** No plant is absolutely deer resistant, as deer will eat whatever they have to in a severe winter, in a drought, or after many plants have died back in the fall. Deer in different regions of the country also have widely disparate tastes in plant matter. Certain families of plants do make deer turn away, however, so it's worth a visit to one of the garden centers near you that specializes in native plants. Choosing plants that do well in your specific climate will help you create and maintain a garden that thrives there, with the added bonus of deer resistance.
    - Plants with spikes, thorns, or prickly surfaces do not interest deer because they are actually painful to eat. Holly, black-

berry, rose, black locust, barberry, Russian olive, and others are essentially deer proof. They also don't like the spiky centers of coneflowers.
- Fuzzy or furry leaved plants turn off deer's tastebuds. Lamb's ear, rose campion, dusty miller, and other plants with softly textured "gray" leaves look beautiful in a garden without drawing interest from deer.
- Plants with strong, spicy, or minty scents like spicebush, prickly ash, lavender, salvia, marigold, lilac, Russian sage, bee balm, catmint, peony, and bayberry are unappetizing to deer, who find the scents distasteful. If you're looking to plant a vegetable garden, choose onions, garlic, mint, and pungent herbs like sage, tarragon, thyme, and rosemary to protect the patch.
- Tough, chewy plants like beech, red osier dogwood, and viburnum are too difficult for deer to consume and digest.
- Flowers like columbine, poppies, foxglove, and daffodils are actually toxic to animals, so deer leave them alone.
- Many other plants and trees just don't appeal to browsing deer: redbud, juniper, serviceberry, spruce, pine, willows, winterberry, and tall flowers including goldenrod, rose mallow, and iris. You can create a stunningly beautiful garden using these and other deer-resistant plants and enjoy it throughout the blooming seasons without the invasion of hungry ungulates.

The little brown bat can slip through a tiny hole in a roof or vent easily, making itself at home in a residential attic or eaves. © GKUCHERA, ISTOCK

Chapter 6

AT HOME WITH BATS

There's nothing quite as satisfying as curling up in bed with a good book at the end of the day, as I did late one spring evening. Our bedroom is on the finished third floor of our American Foursquare home in an older section of a Western New York city. I settled in against the pillows for a pleasant wind down from a busy writing day, looking forward to a retreat into fiction with Donna Tartt's *The Goldfinch*.

I'd barely read a paragraph before something darted past me through the air. I froze, put the book down, glanced around, and saw it again: black, much bigger than a fly, and whizzing back and forth in a fairly obvious panic. In its next pass, its wing grazed my nose.

I am not proud of this, but I dove under the covers and threw the comforter over my head. Reaching out through the dark, I felt for my phone on the nightstand and dragged it under the protective layers of bedclothes. "Call Nic," I whispered to Siri.

Nic, two floors away in the living room, answered, his voice tinged with concern. "Hello?" he said.

"There's a bat in here," I hissed.

"A what?"

"A BAT, for the love of God. You have to help me."

Nic snorted, then managed to stifle any further urge to laugh at me. "Aww, are you scared?"

"Of course I'm scared. Come up here and rescue me."

Nic knew that I have just one animal phobia, and it's an irrational one: I have no issue with bats out in the wild, but one in the house sends me into blind, white-hot panic. I can't explain this—I don't believe the old folk tales about bats getting caught in women's hair, and I know that they are not coming to suck the blood from my neck. I don't even fear that one might be rabid, as this is rare enough in our region to be inconsequential. I just don't want a flying mammal in my bedroom.

Nic, my savior, climbed the two flights of stairs to the bedroom and stood still to see where the little critter was headed. As I peered out from under the covers, I could see that the bat had stopped flying and now clung to the window screen on the west side of the room, perhaps attracted by the streetlights outside. This made its eviction a relatively easy matter: Nic simply removed the screen from the window with the probably terrified bat gripping it in its claws, turned it around so the bat was on the outside, and gave the screen a flick with his thumb and forefinger. It let go and flew off, sustaining its own existence and restoring peace to my bedroom.

Couldn't I have done that myself? I now wondered as I emerged from my makeshift cocoon of sheets and blankets. *Why am I so afraid of a little bat?*

Three days later, I had the chance to redeem myself. When another bat sailed across the living room just after dark, I steeled myself against the adrenaline spike of terror and stood alongside Nic in the foyer, the screen door propped open and a broom in my hand. The bat darted past us again and again, but in a feat that still makes me gasp a bit with amazement, I managed to connect with the flying beastie at exactly the right second, pushing it out the door on the broom bristles and directing it to freedom. It disappeared into the darkness and out of my life.

Two bats in three days didn't bode well, however—and two became three in short order. We arrived home from a weekend trip the following Monday to find a small bat laying quietly on a stair tread halfway up to our second floor. Nic covered it with a plastic bowl and slid a piece of cardboard under it, and we were both relieved to see it move—it was alive and, we guessed, most likely dehydrated. (The possibility that it may have been in the last stages of rabies did not occur to us, but neither of us touched it at all, so further investigation was not required.) He set it outside on a bush still dripping from rain that had fallen an hour before, and in a few minutes, it flew off, leaving only an easily-cleaned-up drop of guano to remind us of its visit.

The third time's the charm, as the saying goes, so the next day I called a pest control company to see if we could discover why bats had made so many appearances in our little house in the space of a week.

The critters scared the bejesus out of me, but I bore them no ill will—they were probably as confused as I was about how they had ended up inside our house. I planned to interview companies to find one that used a humane method of discovering how and where the bats had entered, removing any others that may have taken refuge behind our walls, and sealing them out permanently.

Luckily, the first company I talked with had just such a solution. The technician explained that little brown bats—the most likely species to be in my house, based on my description—often nest inside the walls or attic of human homes, taking shelter against weather while they produce and raise their young. Bats in more southerly climates form nursery roosts in caves, but here in New York, caves are often too cold even in summer to accommodate the bats and their hairless pups, so barns, bat houses, and human homes often become their best option. When the young bats are ready to fly, they vacate the home overnight. In our region, that would most likely take place toward the end of June (this was mid-May).

"We want to protect the bats, because we've lost so many to an infection called white nose syndrome," the technician told us. He explained that this fungal infection spreads between bats living together in caves or other close quarters, and while it doesn't actually make the bats sick, it disturbs their rest during hibernation. Infected bats wake up repeatedly all winter, burning energy that they should be conserving over roughly six months of sleep. The result: millions of bats die every winter. In New York State, where the syndrome was first discovered in Schoharie County at the northern side of the Catskills, the infection has pushed the little brown bat, in particular, into steep decline. White nose syndrome has been detected in bats in forty states and nine of Canada's ten provinces, killing 90 percent of little brown, northern long-eared, and tricolored bats in fewer than ten years.

None of us wanted to endanger the baby bats, so the company would find the tiny openings in the eaves that the bats had used to enter the house, make note of their location, and then wait for another thirty days until the young had matured enough to fly of their own accord.

After thirty days, the technicians returned and installed small, one-way gates in the openings they had earmarked a month earlier. The gates

If bats move into your home, it can take some time to get them out as you wait for their young to reach maturity. © BENJAMIN VELAZQUEZ, ISTOCK

ensured that all the bats would be able to escape from the house when they were ready, but they would be barred from reentry. Sure enough, within two days the occasional rustlings I heard in the walls just behind my head as I lay in bed (oh my God) had dwindled to silence. The bats had taken their leave, and my house was no longer a nesting site; nor would it be a potential hibernaculum for overwintering bats when the weather changed.

While I'm not likely to overcome my visceral aversion to bats in my bedroom, Nic and I made a trip to Carlsbad Caverns National Park not long after the home incident to watch the massive exodus of Mexican free-tailed bats from the main cave at dusk (actually, as many as sixteen other bat species are also represented here, though even our best birding binoculars would not help us identify them in the dark). The ranger who led the viewing program at the mouth of the cave told us, however, that when the bats flew out the previous night, owls had descended on them and caught a number of them for their own consumption. Apparently rattled by this experience, the bats did not come out the night we waited for them. We left at dusk without that checkmark on our bucket list; one day, we will return in hopes of having better luck, and I will find out if witnessing tens of thousands of bats in flight has the same effect on me as one single bat does in my bedroom.

You Already Live with Bats

Forty-seven different bat species inhabit the contiguous United States and Canada, ten of which are on the endangered species list (and another is on the threatened list). Not only do they perish from white nose syndrome, but they also face the same losses of habitat that many other animals experience because of human development. Many bat species live in trees rather than caves, so the more neighborhoods and cities grow and spread into formerly forested areas, the fewer places these bats have to live. The decline of these nocturnal, often unseen animals may be virtually imperceptible to most people, but the result of their absence has the potential to make a significant difference in our quality of life.

Just about every neighborhood in America has its resident population of bats, so whether or not you have actually come face to face with one, you may already enjoy the benefits of having them nearby. Each bat consumes about one thousand insects per hour, devouring as much as 50 percent of its own body weight in bugs in a single night. Without bats, our world would be overrun with mosquitoes, gnats, and other nuisance insects—and if you live in an area where bats have seen major declines, you may already notice an increase in bugs at sunrise and sunset in spring and summer.

Not only do bats keep bugs under control but some also play a critical role in the pollination of as many as three hundred different fruits and nuts. These bats love the nectar in the night-blooming flowers of these plants, so when they move from blossom to bell-shaped blossom, the pollen clings to their fur and drops off in subsequent flowers. Nuts, figs, agaves, bananas, avocados, mangoes, guavas, and coconuts all thrive in part because of bats—and when bats eat the fruits of some of these plants, they defecate the seeds in new places, allowing new plants to germinate. We can be especially grateful to bats around the world for distributing cacao seeds, the main ingredient in chocolate, helping to guarantee a perpetual supply of one of the planet's favorite treats.

As if being the world's only flying mammal (flying squirrels glide, but don't actually fly) were not enough, some bat species have proven themselves to be remarkably adaptable to human-dominated areas. We know that bats communicate using a technique called echolocation, determining their distance from objects by emitting low sounds and listening to the sound waves as they are deflected back to them. This ability can have an important impact on bats' chances for survival in cities, where collisions with buildings could easily be fatal. Remarkably, it seems that some bats have evolved to do well in exactly this kind of environment: scientists at the Leibniz Institute for Zoo and Wildlife Research in Germany have discovered that bats living in cities use echolocation calls at lower frequencies and for longer duration. The researchers also found that urban-dwelling bats showed more flexibility in choosing a day roost, giving them the ability to simply switch to a new place if humans began to move into their first choice.

This adaptability, particularly in smaller bats, makes urban dwelling much easier for the little brown bat and others of its relative size. Even when trees disappear and buildings replace them, these bats can make their home in an abandoned building, an attic rarely accessed by its human owners, or under a large bridge. In Austin, Texas, for example, the largest urban bat colony in the world collects under the Ann W. Richards Congress Avenue Bridge. Experts estimate that from April through October, somewhere between 750,000 and 1.5 million Mexican free-tailed bats fly out from under the bridge every night at dusk and head east, swarming over Lady Bird Lake. The steady stream of bats can take forty-five minutes from its sunset start until it's too dark to see them. This natural phenomenon has become so popular in Austin that the city constructed the Statesman Bat Observation Center adjacent to the bridge, though many people choose to stand on the Butler Hike and Bike Trail alongside Lady Bird Lake. If you want to leave the transportation to others, two companies provide tour riverboats that motor out to the middle of the lake for spectacular viewing. Once they disperse into the night, the bats provide yet another service: they devour somewhere around thirty thousand pounds of insects overnight before returning to the shelter of the Congress Avenue Bridge as the sky begins to lighten near dawn.

At a long-abandoned warehouse at Huntsville State Prison in Texas, Mexican free-tailed bats have formed a maternity roost that contains about 1.2 million individual bats during breeding season. The prison, built in the 1800s, houses inmates who receive training in trades that they can use when they are released, like machinery repair, agriculture, and various crafts. At one time, cotton harvesting served as a training program, with the cotton stored and processed in this building—but after this program ended, bats began moving into the vacant warehouse in the 1990s. A massive fire in the early 2000s did not deter them, even though it left the building's interior in ruins. If anything, the charred remains made the warehouse even more appealing.

More and more bats arrived and joined the colony each year, creating a situation beyond the control of the Texas Department of Criminal Justice (TDCJ). The masonry frame that survived the fire was never meant to support the weight of hundreds of thousands of bats, however, so despite growing popularity of the site as a tourist attraction to witness the nightly bat flight, the colony created one potential health hazard after another.

At first, TDCJ sought to have the bats removed from the building so the remaining structure could be destroyed—but that would mean releasing hundreds of thousands of bats on the unsuspecting public all at once. Where would such a gargantuan colony go? "You don't want to solve one problem and create a giant health emergency for the rest of the area," said Richard Heilbrun, program director for Texas Parks and Wildlife, and one of the department's consultants on the Huntsville roost. Mexican free-tailed bats are a protected species, he said. "If they just tear down the warehouse, they will kill the bats. But bats in an uninhabited building can't be removed." Protected species can't be disturbed—they must be encouraged to leave on their own.

TDCJ has formed a coalition to address the issue, involving Texas Parks and Wildlife, Bat Conservation International (BCI), Apex Clean Energy, and the Texas A&M AgriLife Extension. The group continues to look for "a solution that will benefit the Huntsville bats and benefit the community," BCI notes on its website, "either in the existing building or in a new home nearby. The work is ongoing."

One of the attempts at a solution involved construction of structures on stilts across from the warehouse, each containing enough roof structure to hold thousands of bats hanging upside-down throughout the day. Despite efforts to coax the bats into checking these out, however, the bats essentially ignored them.

The plan announced in 2024 involves replacing the building's roof to make it more structurally sound and then starting a systematic process of closing off areas of the warehouse to make them impervious to bats. TDCJ's Amanda Hernandez, director of communications for the prison, told Houston's KPRC-TV that this process is expected to keep many of the bats from migrating back to the building each spring. "There's definitely some concern about the structural integrity of the warehouse," she said. "It's making sure that we figure out how to do it well and how to do it right to make sure that the bats are safe."

In the meantime, bat enthusiasts can watch clouds of bats vacate the building on most nights from mid-February through September, from the corner of Fourteenth Street and Avenue I in Huntsville.

Capitalizing on the presence of a massive bat colony and on the public's fascination with these mysterious nocturnal creatures can be a good thing for a community. It's easy to understand that the more people know about virtually any animal, the more we all can comprehend why this contributor to environmental harmony should be protected and supported.

This may not be the case, however, for people who find that bats have selected their home as a comfortable shelter in which they can raise their own pups. Many bats return to the same birthing sites and roosts year after year, so what begins as a curiosity may actually turn into a smelly, dirty, potentially dangerous nightmare.

Such was the case for a young couple in the Puget Sound area of Washington State, as reported in the *Washington Post* in October 2024. The couple soon discovered that their attic had been home to thousands of bats over a very long time period, and their walls were jammed with decades of accumulated bat carcasses and skeletons. Living bats continued to raise their young in the attic. A series of pest control experts helped them understand the scope of the problem, but it would take tens of thousands of dollars and hundreds of hours of their own work in full-body hazmat suits to clean out the infestation, close all the gaps in the roof and attic, and repair the extensive damage to the walls. Insurance did not cover the bat remediation because it was a "preexisting condition." Even the precautionary rabies vaccines the family received had to be paid for out-of-pocket.

For all the contributions bats make to our ecosystems, they quickly become the enemy when they move into our homes with us. Bat guano dust can affect the human respiratory system; when it dries and becomes airborne, it releases fungal spores that can spread *Histoplasma capsulatum*, a flu-like illness. (This is usually not fatal, but it can lead to death in immunocompromised people.) Just 1 percent of all bats may have rabies, but most cases of rabies in humans come from contact with bats. If someone in your household is bitten by a bat, do your best to capture the bat alive to take it to a laboratory for testing.

HOW TO SHARE YOUR NEIGHBORHOOD (BUT NOT YOUR HOUSE) WITH BATS

We can live alongside bats without having to share our homes with them, of course. These tips come to us through the nonprofit Bat Conservation International (BCI), the International Association of Certified Home Inspectors, and advisories from several states' departments of natural resources. Much of this advice also applies to attracting birds to your yard, so I've noted that for you.

- **Invite bats over**. If you are hoping to observe bats in your yard or neighborhood, determine which bat species live in your region and create a food supply and habitat that will attract them. Native plants that attract moths and other pollinators will provide a summer-long bug buffet for bats (and birds). Do not treat these plants with pesticides, as this will defeat the purpose of having them by killing the insects. Avoid chemical herbicides as well to keep from poisoning the insects that feed on leaves, stems, and blossoms.

- **If a tree dies, leave it in place**. Dead trees provide important habitat for bats, with loosened bark bats can crawl under and hollowed-out branches and trunk for roosting. If the tree is not poised to fall on your house or garage, leave it where it stands and let bats, birds, and other critters make a winter home in it. The tree will decay eventually and require removal, but you have the additional option of turning it into a pile of logs and branches, where insects can thrive and become food for bats and birds.

- **Provide water.** Water features like birdbaths and automatic drippers attract bats and birds. Change the water every two or three days to keep it clean and fresh. If necessary, use commercially available mosquito tablets to ward off larvae that can develop in standing water.
- **Point lights downward.** To observe bats, you need a dark yard and sky. Turn off your outdoor lights or, at the very least, point them toward the ground so they light paths and sidewalks without ruining your night vision.
- **Keep your cats indoors.** Cats have already been cited as the number one killer of birds in North American backyards and neighborhoods, according to the landmark "Three Billion Birds" study published in 2019. Not only have domestic cats killed more than two billion birds since 1970, hastening the birds' dramatic decline, but the carnage doesn't stop with birds: cats allowed to roam free kill at least eighty-six different species of bats around the world, pushing many species into threatened or endangered status—and that's just what we know from the very limited research on the topic. Keeping domestic cats indoors, especially at night, has the potential to save all kinds of wildlife, from birds to bats to small rodents. If you must allow your pet to wander your neighborhood, make sure it's indoors half an hour before sunset until the following morning.
- **Install a bat house.** Build one yourself according to many plans you can find online or buy one from your favorite nature store. (Check reviews online to see which ones attract bats reliably, as the construction of these can be tricky, and bungled features can actually repel bats rather than attract them.) Bat Conservation International offers a free guide, *The Bat House Builder's Handbook*, a comprehensive reference for building and installing a number of differently sized and faceted bat houses for specific common species: https://batweek.org/wp-content/uploads/2018/01/BHBuildersHdbk13_Online.pdf.

If you are concerned about bats taking up residence in your attic or walls, you have a number of options for preventing this:

- **Plug every hole you can find.** If there's no indication that you have bats in your house, make sure you prevent them from entering by finding any openings they might use to get in. Bats are generally much smaller than you might think after seeing one with outstretched wings. Some bats can fit through openings as small as half an inch wide, making it very difficult to keep them out of your house without a thorough inspection of the roof, eaves, and walls. Bats look for openings in high places, as they need a swooping descent to begin to fly, so examine the highest areas of your home (or hire a pest control company with tall ladders to do so).
- **Know the signs.** If you see any of these indications of animal activity, you may have bats in your home:
    - white stains on windows from bat urine;
    - tiny holes and cracks with white stains around them;
    - droppings similar to the pellet-like leavings of mice, collecting under eaves or on the floor of your attic;
    - guano odors (ammonia and urine smells);
    - noise in your walls: scratching, squeaking, skittering claws; and/or
    - smeared dirt or grease around the edges of an opening.
- **Get an inspection.** Call a pest control professional to help you find all of the places where bats may be entering your home. If the bats' residency in your house is new and only a few bats have gained entry, you may be able to plug the holes yourself after the pups are mature enough to fly (in most of the United States, this will be in late July/early August). Depending on the extent of the damage inside your house (if any), you may have a difficult job on your double-gloved hands to remove the guano and make repairs.

One of the best and most detailed checklists for this can be found on the Government of Quebec's website at https://www.quebec.ca/en/housing-territory/healthy-living-environment/cleaning-of-an-environment-contaminated-with-bat-droppings. If you have the resources to hire a professional to do this, by all means do so—and if you have a larger infestation, you may have no choice but to work with a pest control company.

All of the contiguous forty-eight states and Alaska have coyote populations, as these adaptable animals can make a living almost anywhere. @ NIC MINETOR

Chapter 7

TOP OF THE FOOD CHAIN

COYOTES

When Christopher Nagy walks his dog in a New York City park, he diligently picks up poop—but not just from his own dog. As a wildlife biologist and the cofounder of the Gotham Coyote Project, he has made a career of collecting coyote scat in parkland and open spaces across the nation's largest city and bringing it to the American Museum of Natural History in Manhattan for in-depth analysis.

That sounds like a bizarre professional choice, but you haven't heard the best part: He and his cofounder, conservation scientist Mark Weckel, have assembled a team of high school and graduate students who willingly perform the analysis, using tweezers and fine tools to pick bits of bone, fur, and other undigestible remnants out of the scat, under the supervision of the Museum's mammal collections manager, Neil Duncan. Mark's work as the director of youth initiatives at the museum gave him an inside track to engaging young people in this unusual task.

Coyotes (*Canis latrans*) were first seen in New York City in 1994, and Nagy discovered the first breeding groups in 2012—raising young in the middle of a metropolitan area with more than nineteen million human residents. The recent development of this microcosm of urban coyotes gave researchers a unique opportunity to follow the coyotes' progress in this metro area and see how they were adapting to their surroundings and to the available food supply—the animals and plants that they would normally eat in the wild, and the anthropogenic food sources being supplied by humans and their endless supply of garbage.

When Nagy is not in the field, he's the director of research and education at the lovely and walkable Mianus River Gorge in Westchester County, but the gorge's management has helped him make time in his schedule for semi-weekly visits to the New York City boroughs to collect what coyotes leave behind. "We see a majority of our diet in scat," he told me in a Zoom interview. "But when they digest most anthropogenic things, there's nothing macro-sized left to tell us what it was. If a coyote eats a sandwich, like a burger out of the trash, pretty much all the parts of that sandwich get digested and don't leave anything behind."

For a more complete picture, the researchers needed to analyze the scat samples for DNA remnants of all of the food they could detect. They sought a partner for this and engaged Carl Henger, a graduate student at Fordham University, who had an interest in large mammals.

"Instead of physically sifting through the poop for bones and hair, she did a broad-sweep genetic survey of the poop itself," said Nagy. "She could pick up things we could not find otherwise."

So what are the Gotham coyotes eating? Mostly rodents and other mammals, with 34.7 percent of the scat samples containing small mammals like mice and rats, and 31.6 percent testing positive for somewhat larger mammals. Racoons came next, showing up in 27.4 percent of the samples, as did birds—mostly rock pigeons and other birds of city streets.

"They are eating a lot of raccoons," said Nagy, which might make some New York residents more than a little pleased with their coyote neighbors. "Raccoons are pretty big and pretty fierce, but they are just so numerous in the city that they can't not eat them."

On the vegetation side, 80 percent of the samples contained some kind of plant matter, with meadow plants including aster and grass leading the list, though plants in the rose family showed up in 25 percent of the samples. About a third of the coyotes had consumed insects.

That all seems like enough food to satisfy a coyote, but the list goes on: more than 64 percent of the samples contained human foods, with chicken topping the menu in 48 percent of the coyote scat. Of the wide variety of other foods coyotes might be able to acquire in the city, only pig products (ham, pork, or bacon) registered any kind of preference, showing up in about 18 percent of the scat.

In results of this study published in 2020 in the *Journal of Urban Ecology*, the scat collected in Pelham Bay Park, the largest Bronx park in the study, actually contained about 19 percent deer remains, indicating that coyotes there maintained at least some of their natural attraction to traditional food sources. At the other end of the spectrum, the coyotes in a small green space in the Elmjack neighborhood contained the most human food and general debris, in a congested area where some tree clearing and construction began on the site during the study.

None of the 146 coyote scats collected contained any kind of dog and only 4.4 percent contained cat remains, dispelling the belief that coyotes thrive in cities by eating pets. While coyotes are well known to be opportunistic feeders—catching and devouring whatever food item is the shortest route to a meal—small dogs or cats with human companions probably do not look like easy prey.

So, coyotes in New York City do not depend solely on people's cast-offs or their pets for their livelihood, choosing the foods their instincts lead them to prefer over what they might find in trash cans or dumps, or on city streets.

A 2022 study expands on this research, finding that the scat from coyotes in Rockland and Westchester counties north of the New York metropolitan area revealed a marked preference for raccoon, rabbit, meadow vole, and pigeon. Urban coyotes, however, demonstrated a taste for chicken—in fact, 48.4 percent of their scat samples contained some of this. They also ate more plant matter than their nonurban counterparts, favoring aster, rose, grape, and grass. Again, just 4 percent of the urban animals' scat contained traces of domestic cat, and none had any evidence that they had eaten dogs.

This all brings us to the third point mentioned earlier in this chapter: People tend to warm to animals in their neighborhood when they see them as somehow useful, either as fascinating beings that provide hours of pleasant observation, or as actually doing something we need.

Coyotes have made themselves very much at home in hundreds of cities from New York to Chicago, Dallas, and Los Angeles. © JIM GLAB, ISTOCK

Many birds, for example, spread seed to grow more plants by eating berries from a fruit-bearing plant like a holly or spicebush and defecating the remaining seeds somewhere else. Bees pollinate our flowers, keeping our plants healthy and producing more blooms. And coyotes eat rodents, making a limited but positive contribution to solving our cities' pest problems.

It may seem that coyotes' preference for rodents might be a boon for New York, where the government recently declared all-out war against rats. The rat population explosion does not actually offer much to other animals in the city, however, because of one of the solutions already in practice: poisoning the rats. The poisons remain in the rats' tissue and digestive system after death, so any other animal that eats a contaminated rat will get sick and probably die.

"We really don't want coyotes eating poisoned rats," said Nagy. "It's a serious thing."

Luckily, coyotes don't spend a lot of time sniffing out rats on city streets. For the most part, New York's coyotes stay out of sight as much as they can, living in places like Central Park and doing their best to avoid the hundred thousand people or more who cross through that park on a daily basis. "We have seen them in Central Park attempting to mate," said Nagy, "but they need a den that's fairly secluded. There are a lot of parks that are smaller than Central Park than have fewer people visiting, and the coyotes in those parks have pups. So once they have food and vegetative cover, how much human activity will they tolerate? My theory is that they are more sensitive to this if they have a den. If it's too much, they will leave the den and not come back to that spot. I think there are places in the city where they've done this and it works out, and others not so much."

Nagy sometimes assists the city's park service when they get calls about coyotes that residents consider a problem—animals that frequent their yard too often for their comfort, or that take the blame for the occasional disappearance of a small pet. He hopes that over time, the Gotham Coyote Project can play a role in reducing the number of people who see coyotes as a nuisance. "The idea is that we live in nature and are of nature, and we may have animals around us that are a little larger than we expect," he said. "They are still wild animals; they are not your pets. People often go one of two ways: either they hate them, they're scared of them, and they think we should kill them all, or people want to entice them to come closer, luring them in with a sandwich so they can get a selfie. Neither of these are good things."

We should take a lesson from the birdwatching community, he added. "Birders do a good job of getting people to enjoy nature. I'd like to have that happen for coyotes in the city, too."

THE SCARY SIDE

With a simple search on "coyote encounters," stacks of videos come to light on YouTube, but they show us a different perspective on living and exploring in proximity to animals that are accustomed to human presence. Very separate from both wolves and dogs, coyotes are dog-like enough to make us drop our guard around them, giving them an unintended opportunity to carry off a small cat or dog, or even to grab a human child if they were inclined to do so. Indeed, as if to prey on our fears, most of the coyote videos that make it to social media involve the animal actively pursuing a cat, small dog, or even a toddler as prey, running fast to overtake and overpower its target and grabbing the child or pet in its jaws. Nothing about the encounter is subject to interpretation. These animals mean business.

If just a couple of videos looked like this, we could dismiss them as unusual cases. The sheer quantity of them, however, makes the concept of sharing our neighborhoods with coyotes more ominous than we might expect. In one, a coyote runs up and attacks a toddler on a California beach, mauling her right behind her parents' backs for agonizing seconds until they finally turn around and rescue her. News reports tell us that the child had severe but not life-threatening injuries.

In another, a man in a bathrobe intercedes when a coyote attacks his small dog in his own backyard, grabbing the coyote by the tail and throwing it into what appears to be his compost bin. He snaps the lip down on top of it, still hollering at his own dog to "git." (The dog seems to be unhurt.)

Many more of these present equally frightening situations: a child running past her father in her backyard, yelling, "There's a coyote!" with the animal close enough behind her to nip at her heels; a toddler attacked on the sidewalk as she waits for her father to get out of the car, with the coyote attempting to drag her down the street; a domestic cat fighting for its life on its own back porch and finally climbing up a post, clinging near the roof until the coyote gives up and wanders off. It's enough to make every homeowner want to erect ten-foot fences around their yards to keep human life in and wildlife out.

As highly publicized as these encounters are, experts say they are very much the exception, not the rule.

"Coyotes get a bum rap," said Stanley Gehrt in his 2024 book *Coyotes Among Us*. Gehrt is chair of the Center for Wildlife Research at the Max McGraw Wildlife Foundation northwest of Chicago, Illinois, and principal investigator of the Urban Coyote Research Project since its outset in 2000. "They are almost never noticed until they are perceived as a problem."

We may see a dozen or more attacks by coyotes in videos and news reports across the country, but they are newsworthy because they are so uncommon—and viewing all of them at once deceives us into believing that they are much more frequent than they are.

Generally, coyotes blend so thoroughly into the landscape of metropolitan areas that we don't even know they are there. City parks and green spaces may each have their own coyote pack, or if the spaces are fairly close together or connected by bridges and open land, one pack may hold sway over an entire area. Either way, coyotes are definitely there, tending to their own lives and keeping the rodent and pigeon populations at bay—one of the principal benefits of having these wild neighbors.

In most cities, coyotes are the apex predator (just below humans), seeing virtually all the other animals in their territory as potential meals. Rats, rabbits, pigeons, voles, moles, groundhogs, squirrels, chipmunks, deer, and more all fall into this general category of sustenance—but so do domestic animals, as coyotes have no way to understand the concept of pets or their relationships with people. This means that they run afoul of humans when they choose a small dog or cat as a potential meal. In such a situation, a coyote can move from being a fascinating curiosity to a bloodthirsty enemy, one that seems likely to return to try again if it does not get the meal it attempted to catch.

So how do we live alongside animals that hide their presence so well, and that are particularly adept at emerging out of nowhere, dashing past, and grabbing what they want? Coyotes present a different kind of challenge, one that requires us to form a mutual respect born of wariness.

A Different Definition of Safe

As dwellers at the top of the food chain, we humans tend to view animals in our urban and suburban areas through several very human points of view:

1. Is this animal going to hurt me, my family, my pets, or my property?
2. Does this animal like what my property offers, and does it then "like" me?
3. How is this animal useful to me, my family, or my property?

The first is a valid and appropriate concern in most cases. We build homes and create boundaries in yards in part to keep ourselves and our families safe, so visits by an animal we don't know well can seem intrusive and even frightening. This holds true even more when the animal is a coyote.

Our love of anthropomorphizing animals—attributing human characteristics, motivations, and emotions to them—can make us even more wary of the creatures that visit our property. It's one thing to imagine that a black-capped chickadee at a bird feeder appreciates the offer of sunflower seeds and extends that appreciation to us, somehow developing a genuine affection for the people who keep the feeder full. It's quite another, however, to leap to the conclusion that the chickadee is sitting and waiting for a human to fill the feeder so the bird can peck the person's eyes out in an act of pure spite. Chickadees don't plot against us; they haven't the mental capacity for that.

It's the same kind of stretch when someone decides that the first-time appearance of a coyote in their neighborhood represents some kind of evil intent on the coyote's part—as if its very existence puts all of the children and pets in the area at risk. It doesn't help that a 1950s children's television show represented a coyote as an obsessive predator—though Wile E. Coyote was completely inept at catching so much as a tailfeather of his nemesis, the Roadrunner. (For the record, a greater roadrunner's maximum running speed is about twenty-six miles per hour, while a coyote can run at speeds beyond forty miles per hour. So, Wile E. lost out because of his lack of brains, not brawn.)

The third point, however, changes our perspective entirely: If we can find some benefit to human residents in having a coyote family in our neighborhood, the animals' presence becomes acceptable, or even desirable. "People are always looking for ecosystem benefits," said Nagy in our interview, "but overall, wildlife have their own benefits," including their voracious appetite for the vermin living among cities' trash.

This brings us to the Urban Coyote Research Project, an ongoing initiative centered in Chicago—the third most populated city in the United States, with more than 2.7 million people in residence. Taking the entire metropolitan area (Chicago-Naperville-Elgin) into account brings the population up to 9.26 million as of 2023. Distributed throughout this area are between two thousand and four thousand coyotes, according to Gehrt.

In fact, Chicago is not at all unusual in the number of coyotes that live within its metro area. In *Coyotes Among Us*, Gehrt and coauthor Kerry Luft notes that while many coyotes continue to live in the Western deserts and the Great Plains, they now inhabit forty-nine of the fifty states (all but Hawai'i), adapting readily to mountain, bayou, and river delta landscapes. In the 1990s, they began to move into suburbs and cities, somehow evading traffic, traps, domestic dogs, and other obstacles to become part of the urban landscape. In addition to Chicago and all five boroughs of New York City, city residents in Los Angeles and Seattle see them walking through neighborhoods and lounging in parks.

The Urban Coyote Research Project seeks to learn all it can about coyotes in cities, both to inform humans about any potential threats these animals may have on them or their environment and to keep conflicts between animals and humans to a minimum. In their decades of research, one thing has become abundantly clear: coyotes are one of the most resilient and determined species on the continent; they are not going anywhere, so it's imperative that we learn to live alongside them.

How We Got Here

There was a time in North America when the name Coyote meant more than the furry four-legged animal with whom Indigenous cultures shared the land. Among the many tribes in what is now California, a character known as Coyote—a powerful, anthropomorphized version of the familiar animal, usually described as standing upright on two legs—actually had a hand in creating the Earth itself.

Coyote and an ethereal being called Earth Maker sang together to bring the world into being, but once it existed and Earth Maker had populated it with animals and people, Coyote turned on Earth Maker and attempted to ruin the world with dirty tricks and evil. Earth Maker and his new legions of people did their best to foil Coyote and try to wipe him out of existence, but they could not do it; Coyote was every bit as clever and powerful as Earth Maker. The two eventually had to find a way to coexist.

This ancient story—one of many about Coyote in California and across the Western United States—seems like a parable for what actually did happen to coyotes after Europeans arrived and began to move across the continent. We have seen in earlier chapters how homesteaders in the Great Plains and farther west fought wolves, bobcats, and mountain lions to protect their livestock. They also turned against coyotes that stole their young lambs, calves, and chickens, though the smaller animal with the appearance of a large dog did not seem as threatening to people as wolves and cougars. Faced with human aggression, coyotes employed their stealth and their willingness to roam, expanding their range to wherever they could find a steady food supply and avoid human interaction. It didn't require much travel to dodge humans in the nineteenth and early twentieth centuries, so coyotes had plenty of land to sustain them, and all kinds of small wild animals to keep their bellies full.

As transportation methods improved, however, people established towns and cities along virtually every railroad route and river, eventually building highways to replace the crude roads that crossed the country. The human population exploded in North America, infringing on the wide-open spaces coyotes had shared only with other animals. Even though humans had done coyotes a major favor in extirpating wolves and cougars—the only two animals that sometimes preyed on coyotes—the coyotes still had to adapt to this new predator in their midst.

And adapt they did, in ways that have baffled wildlife science professionals for decades. Instead of receding from land taken over by housing developments, office buildings, and parking lots, coyotes moved into some of the most human-populated areas on the continent. By the 1960s, they had begun to drift steadily east from their accustomed territory, showing up in the rice fields and bayous of Louisiana and Mississippi and in the forests of northern New England. They pushed into Florida and the Carolinas by the 1990s, and as the millennium ended, they reached every one of the contiguous forty-eight states and well into Alaska. Today coyotes can be spotted throughout Central America and in Canada's northern provinces.

"The amazing aspect of this tremendous range expansion is that it has been accomplished in the face of ongoing and intense human persecution," Gehrt notes in his book. He points out that the coyote has never been a protected species; humans have hunted them for hundreds of years, with no limits in most states on the number of coyotes they can kill. The annual take by coyote hunters in the United States alone approaches one million animals—and that doesn't include the thousands that die in vehicle strikes.

How did coyotes accomplish this expansion? The method has a great deal to do with the social system that scientists have observed through coyote study projects in Portland, Oregon; Madison, Wisconsin; Edmonton, Alberta, Canada; Los Angeles and areas of northern California; and Bronx, New York. Coyotes live and move in family groups, or packs, choosing a territory and defending it from all coyotes outside of their unit. As their young mature, they begin to look for their own territory to bond with a mate and raise a family, but as these packs fill a given area, the young adult coyotes must go farther and farther from their place of origin to establish their own stronghold. This process was not so difficult in the vastness of the Great Plains or the Southwestern deserts, but in a bustling American city, it means staking a claim in a human-built subdivision, in a park not already claimed by another pack, in a patch of woodland between properties, or even in a local cemetery. In their chosen habitat, coyotes make themselves invisible by day, taking cover under some ground-level foliage or in the shadows between bushes, and emerge at dusk to pursue their prey. Often, they sit or lie so still that people pass by them and never know that there's a coyote just a few feet away.

What does this mean for homeowners? If you wish you had fewer squirrels at your bird feeders or if your compost has attracted the wrong kind of rodent, your friendly neighborhood coyote pack may be just what you need to rectify the situation. Think of them the way you do the neighbor who blows the snow out of your driveway after the plows have come through—a cohabitant doing a favor and wanting nothing in return but a peaceful environment.

It's interesting to note that coyotes and feral cats have developed an understanding of sorts, giving one another wide berth in the wild. These are not house cats with a propensity for the outdoors; instead, feral cats are unadoptable, probably living rough since they were kittens and unable to adapt to living under the care of humans. As they wander on their own or in groups, they deliberately avoid natural areas that probably contain coyotes and other truly wild animals. This protects them from predation; in fact, in a recent study, the Urban Coyote Research Project outfitted more than 120 feral cats with radio collars to monitor their whereabouts and found that just 7 percent of these cats fell prey to coyotes, even though they shared the same metro area.

When to Reach Out for Help

Most coyotes want nothing to do with people, so seeing one near your property does not mean that it's going to come after you, your children, or your pets. A coyote behaving just like a coyote—only showing up between dusk and dawn and shying away from you when it sees you or when you make noise—will simply pass through your area or go about its business while paying no attention to you.

If a coyote begins to pay your yard regular visits, however, it may have developed a comfort level around humans that will not serve it well. You can test this by hazing the animal to see if it will scare off. If it does not, it's time to watch this coyote and see what it intends to do. Stalking your pets, coming close to your house or porch, eating your trash or compost, or otherwise raiding the human food supply can all be signs of a habituated animal. Now it's time to alert your local wildlife professionals to the coyote's whereabouts and behavior.

How to Share Your Neighborhood with Coyotes

- **Don't feed the coyotes**. No matter how many videos you see on YouTube of people feeding coyotes from their back deck, this is a very bad idea. Just as with any other wild animal, luring coyotes to your porch with a package of hot dogs or a bowl of pet food may be fun for you for a time, but it does nothing to help the coyote. Animals that learn to eat what humans provide for them become habituated to interacting with people. That makes them dangerous, as they will approach other people to beg for food and may become aggressive when they don't get handouts. Coyotes are expert hunters and scavengers, surviving cold winters and other adverse conditions without human assistance. Let them find their own food without your "help."
- **Remove temptations**. While coyotes feed primarily on small, wild rodents, their diet can change if other food items become available to them on a regular basis. Studies mentioned earlier in this chapter have shown that coyotes living in urban areas do not rely on human castoffs, garbage, or pets as their diet—they focus on rabbits, birds, mice, rats, white-tailed deer, and fruit, except in seasons where these dietary staples become scarce. In winter, for example, coyotes may take more of an interest in the contents of trash cans to assuage their hunger. Keep trash cans out of sight except on pick-up day, even in warmer seasons when coyotes are less likely to plunder them.
- **Lock up your compost**. Coyotes will dig through loose compost to find food scraps like fruit peels, discarded vegetables, and processed foods. Use a compost bin with a cover that you can lock to keep these and other animals out of the pile.
- **Take down bird feeders**. If you know you have coyotes visiting your yard or neighborhood, remove your feeders so coyotes won't use them to prey on the birds and small rodents that eat the seed. Coyotes will also eat bird food if they can reach it.

- **Keep your pets and pet food indoors.** Domestic pets are not a mainstay in a coyote's diet, but these opportunistic feeders may go after a small cat that comes their way if they're not finding a more conventional meal. When it comes to encounters with wildlife, it's in your animals' best interest to keep them indoors and out of harm's way. Always walk your dog on a leash; if you encounter a coyote or a pack on your walk, maintain control of the dog until you and your pet have passed by.
- **Hazing may help.** Repellent behavior—banging pots together, flashing lights, using motion-activated devices like noisemakers—has some limited success with coyotes. If you encounter a coyote in your yard or on a trail, waving your arms and shouting usually is enough to ward them off. If the coyote continues to approach or does not walk away, try throwing a non-food item at it (like a small rock).
- **Fence your yard.** If you live in an area with high wildlife traffic, a fence around your yard may be warranted to keep your garden, bird feeders, and pets safe. An effective fence against coyotes must be at least six feet high—in fact, some sources say ten feet—preferably with a roll bar or chicken wire at the top to keep the animal from scrambling up and gaining enough footing to jump over it. (If you are also fencing against deer, you'll need a fence at least eight feet high, which will also foil the coyotes.) If there's something in your yard that a coyote really wants (like your trash or food scraps in your compost), it can dig under a fence to get in; to foil this behavior, extend your fence down into the ground at least twenty-four inches.
- **Educate your neighbors.** It only takes one resident with dishes of pet food on the porch to turn a neighborhood into coyote heaven. Coyotes that become habituated to people food also become used to people—increasing their opportunities to test whether humans and their pets could be easy prey. Work with your neighbors to distribute information that discourages these folks from deliberately feeding or attracting wildlife.

- **Don't overreact**. A coyote visiting your yard or neighborhood for the first time is probably just passing through. Do your best not to be alarmed by this, especially if it seems disinterested in humans or anything your yard may have to offer. We share our world with all kinds of animals that want nothing at all to do with us; there's no need to call the authorities if the coyote and its pals are minding their own business.

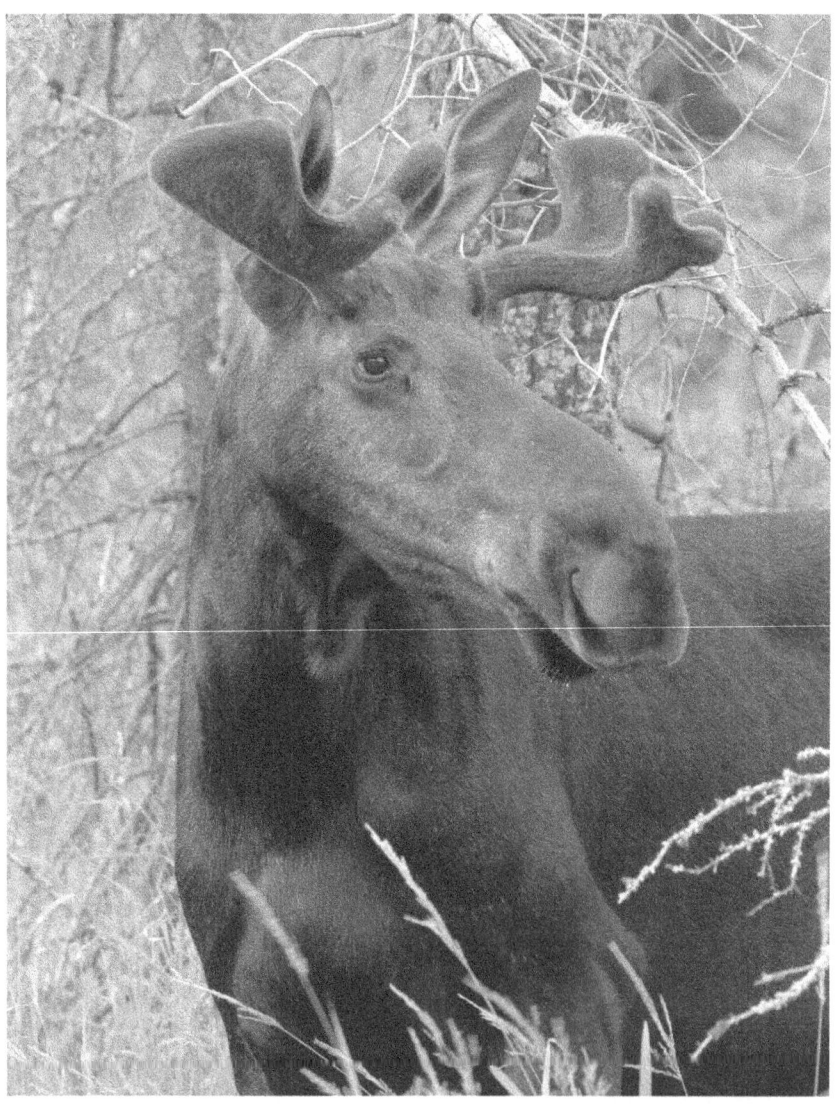
Moose, America's largest member of the deer family, do not seek out residential areas—but sometimes they make a surprise appearance. © DANA J. COFELL, ISTOCK

CHAPTER 8

THE UNEXPECTED MOOSE

VISITORS TO DOWNTOWN SANTA FE, NEW MEXICO, HAVE SPECIFIC, VERY valid expectations as they walk along the town's residential streets: terracotta-colored houses, adobe walls defining the property lines, and xeriscape yards covered in gravel, with yucca and agave plants providing visual interest. On temperate days, they may catch a glimpse of a fox squirrel, a gray squirrel, or even a greater roadrunner living up to its name, as well as the occasional coral snake.

No one expects to see a moose in Santa Fe. Moose are not native to New Mexico, though a few individuals have moved into the state's northern mountains, expanding their range from more frequented habitat in southern Colorado. As long as moose can find their preferred woody plants to browse on, they can make a go of the mountainous terrain in the lower Rocky Mountains, where aspen and willow grow in abundance and wet areas provide marsh plants with the high sodium content moose require. Northern New Mexico's small population of moose arrived in the 1990s, largely staying out of sight of people except for the occasional amble along the edge of a highway. Sightings in Taos, Chama, and Tierra Amarilla are so uncommon that they usually make the local news.

So when a bull moose strode into the middle of Santa Fe one day in September 2023, it caused something of a commotion. This, locals informed curious visitors, was the animal affectionately nicknamed Marty Moose, a lone bull that had hung around on the outskirts of the city for the past year. Marty attracted so much interest that he warranted a now-retired Facebook page titled "Where's Marty Moose?" with reports of each time one of his many fans spotted him in a field or on a highway. Video of him trotting up Hyde Park Road on the way to the Ski Santa Fe resort over the winter ran on all the local news channels.

Most animal families split up as the young grow up and go off to choose their own territories, where they will breed and raise their own young. Over time, this process of finding a new home expands the range of all kinds of animals, from lizards to birds to large ungulates (hooved animals) like moose, bison, and deer. The drive to have an area of his own most likely drove Marty away from his family's accustomed habitat in Colorado—but this far south, with no other moose to be found, Marty would not find a mate or become a father. This loner had wandered into the moose equivalent of exile, finding plenty of food and adequate aquatic plants and water, but never coming upon another of his own kind.

Fascinating and, thus far, harmless, Marty was left to his own devices as he enjoyed the twigs and bark of the New Mexico mountains' ample trees and drank from the rivers in the area. When the emboldened moose decided to explore the space where humans lived, however, wildlife officials agreed that it was time for a relocation.

Visitors and residents of Santa Fe gathered at a respectable distance to keep watch over him until the authorities arrived, taking photos and generally enjoying the unusual moment. City of Santa Fe Animal Services officers arrived on the scene around 8:00 a.m. and called the New Mexico Department of Game and Fish, the bureau best equipped to figure out what to do with the enormous, out-of-place ungulate.

Officers tracked the moose to the intersection of Grant Avenue and Rosario Boulevard and quickly formed a perimeter to protect it from traffic, getting tangled in a fence, or any other potential hazard a moose in a city might encounter. Moose do not go out of their way to attack people, but a charge by this nine-hundred-pound animal could have serious consequences for whoever managed to get in the animal's way. Keeping people at a safe distance became a major priority, just in case this moose decided that the onlookers were infringing on its newfound territory.

Remarkably, Marty seemed nonplussed by the entire operation, going about his usual business of munching on various plants as officers beyond the established perimeter began to plan next steps. The unsuspecting moose finally felt a tranquilizer dart pierce his right flank. In a few minutes, he staggered, fell to his knees, and slept—and a full dozen officers gathered around and lifted the animal into the back of a trailer, grunting under his massive weight.

Before they drove off, wildlife officials took advantage of the opportunity to examine the moose, finding him to be a healthy animal of about four or five years old. They released him in far northern New Mexico—photos of the moment when he left the trailer capture a facial expression not unlike a smile—and hoped that Marty would soon encounter the moose herd known to be living in that area.

"This moose will hopefully find a happy ending to its long journey," officials said in a news release. "We are happy to see a positive outcome for this moose, who can now thrive in quality habitat where it does not pose a threat to public safety."

Moose in Our Midst

Why would a moose decide to stroll into a city? The largest animal in the deer family does not want or need anything that humans have, so its occasional appearance in a human-populated area always feels like a special event. Standing as high as seven feet tall and weighing more than one thousand pounds, a male moose with a full rack of its iconic, palmate (hand-shaped) antlers looks like a creature we should fear, even though most moose are barely distracted from grazing by humans in their presence. Indeed, we should be wary of these enormous animals, even though wildlife specialists note that they are not naturally aggressive toward people—at least, not usually. In fall, a bull moose in the throes of the breeding season can be unpredictable and easily provoked. The smaller, 850-pound females go on the offense quickly if their young are nearby.

Moose feed exclusively on plants, eating readily available leaves and twigs of willow, birch, maple, and aspen trees, as well as fruiting trees like cherry and mountain ash. In Maine, where some seventy-thousand-plus moose live—that's more than in the other forty-seven contiguous states combined—moose switch to the bark of balsam fir trees in winter, augmenting this low-nutrition diet with whatever browse is available above the snow.

They supplement this diet year-round with aquatic plants like pondweed and water lily when icy weather has not frozen the wetlands solid. The water plants provide higher levels of salt than the rest of a moose's diet, so when winter descends, and they can't access these sodium-rich plants, moose may take to the edges of roads to lick up the salt spread by municipal plows. This behavior brings them into contact with humans—or, more specifically, with the vehicles we drive. In the murky light of dawn and dusk, moose on the roadside can place themselves in great peril, blending in with dark mountainsides despite their height and bulk. Striking a moose virtually always results in damage to the car, injury or death to the moose, and injuries to the driver and his or her passengers.

Beyond these appearances along highways, however, moose do not go out of their way to venture into human territory. They have no interest in whatever may be in a household's trash or compost bin and as committed vegetarians, they do not compete with carnivores for rodents. When a moose shows up in someone's yard, its only interest is in their garden or trees.

"When moose wander into lower elevations, there is no more desirable food source than ornamental plants, fruit trees, garden plants, and flowers that people plant around their homes," wrote Phil Cooper of Idaho Fish and Game's Panhandle Region in *The Spokesman-Review*, the newspaper of Spokane, Washington. "Our landscaping invites moose to dine amongst us."

Most people who encounter a moose see one walking along a highway near the appropriate habitat, like a coniferous forest. © GARY GRAY, ISTOCK

Moose visits to suburban yards may be more frequent in spring, when tender new shoots, buds, and leaves emerge and early flowers like tulips look especially tasty after months of tree bark during the cold season. In a year with severe drought, moose may make more frequent forays into human-populated areas to find water sources.

People living in moose country from Maine to Alaska describe moose stripping every twig and leaf off of their apple, cherry, and plum trees before their eyes, a feat the animals can accomplish because they are so much taller than other ungulates. This is why the Indigenous Algonquian people of Northeastern North America named these animals "moosu," which translates as "he who strips off."

People who have waited through a long winter to see their flowers in bloom may not appreciate the intrusion of a moose on their cultivated garden. For those who have invested time, money, and loving care into a vegetable garden or yard filled with fruit trees, the appearance of a moose—an animal that eats one hundred pounds of food every day—can lose its novelty and fascination in minutes and quickly move into heartbreak. In most cases, the moose will eat its fill and walk away, but it may strip your trees and flowers down to the ground before doing so.

Still, given that moose populations across the continent face a mounting struggle as climate change warms their usual habitat, we can't blame them for taking the easiest route they can find to a good meal. These cold weather denizens have not had an easy time for decades, and their history of cohabitation with European settlers has been uncomfortable from the beginning.

Saving Moose from the Brink

Indigenous people who lived in what are now America's Northern states made a good meal of moose regularly, sometimes stalking the huge animal for an entire day or more until it finally fell to their attacks. Moose were plentiful for thousands of years, making their first entrance into North America around fifteen thousand years ago over the Beringia land bridge, the ninety-mile-wide strip of land between northern Asia's eastern and Alaska's western shores before ocean waters submerged it. From here they spread across the continent. Millions of moose roamed through Canada, into the Rocky Mountains, and around the Great Lakes and beyond, finding particularly attractive, heavily wooded country in what would become New England.

When European colonists arrived in the early 1600s, they immediately saw the advantages of moose hunting. One bull moose could yield hundreds of pounds of meat, feeding a family for weeks; a colony's team of hunters could kill enough moose to feed the entire settlement for the winter. Diarists of the late 1600s and early 1700s tell stories of felling twenty moose and caching the meat, keeping it cold by immersing it in a river. As colonies became towns and cities grew, moose appeared plentiful in the heavily forested lands of New England, so regulating the take was not a priority; more hunters simply killed more moose and fed more people.

More people also needed to grow fruit and vegetables and grains for bread, so they cleared land and felled forests, turning the logs into wood to build homes and barns. Forest-dwelling animals were left with less and less habitat, so by the mid-1800s, moose became a rare sight. The killing of a young bull moose in 1899 in Essex County, Vermont, was hailed as "the last moose in Vermont" in the local newspaper. By this time, hunting was well regulated throughout New England, with Massachusetts leading the way in limiting deer and moose takes, but with pressure on the population from all sides, moose disappeared from much of the region.

Maine managed to protect its moose before their numbers diminished, however, and some of its herds eventually moved west and back into Vermont, New Hampshire, Massachusetts, and Connecticut. Other regions took a more active part in restoring moose to their area: in Michigan's Upper Peninsula (UP), an international effort in 1985 and 1987 involving wildlife biologists and veterinarians from the United States and Canada pulled off an operation called the Michigan Moose Lift, transporting fifty-nine moose individually by helicopter from Algonquin Provincial Park in Ontario to a new range west of Marquette, Michigan. A recent recap of this feat on the Michigan Wildlife Council website tells us that there are now ten times that many moose in the UP, in "a core area that includes parts of Marquette, Baraga, and Iron counties," said Dean Beyer Jr., wildlife research biologist with the Michigan Department of Natural Resources, to the anonymous reporter.

Even beyond their attractiveness as a food source for humans, moose face a number of challenges brought about by their environment. They share their habitat with white-tailed deer, which can carry parasitic meningeal worms, or brainworms (*Parelaphostrongylus tenuis*) that have no effect on the deer except to thrive in their feces. When the deer leave their waste in the forest, land snails eat it, ingesting the brain worms and becoming carriers as well. Moose eat the land snails as they graze on trees, releasing the parasite into their own digestive systems, where it finds its way to the moose's brain. Animals infected with this "moose sickness" become weak and uncoordinated, lose their hearing and sight, and eventually become paralyzed and die. Brainworm affects moose through the eastern half of North America, especially in New Brunswick, Nova Scotia, Minnesota, and Maine. One of the keys to the successful growth in the transplanted moose population in Michigan's UP was in making the habitat unattractive to deer. This strategy reduces brainworm cases, easing one of the challenges the moose face in this region.

This predator on its own would be more than enough of a threat to moose survival, but another tiny enemy feeds on the unsuspecting moose. The winter tick (*Dermacentor albipictus*) does not carry diseases the way that deer ticks do, but when thousands of these tiny parasites gang up on a moose's flesh, they devour enough blood to fell the enormous animal. As many as seventy-five thousand of these ticks have been found on a single moose, according to a *National Geographic* account, making them quite impossible to defeat. This phenomenon is a direct result of climate change, as longer summers give winter ticks more time to lay eggs and hatch in late summer, allowing them to reproduce in far greater numbers. Their larvae then attach to the moose as it browses through a meadow of tall plants and develop into their adult stage on the moose's body. By March, the female ticks have filled themselves with the moose's blood and are ready to reproduce.

Well suited to cold climates, moose have an unusual coat that consists of hairs that are hollow, trapping air inside each hair and creating a layer of insulation that allows them to walk through deep snow in relative comfort. This makes them vulnerable to the higher temperatures brought by climate change. Prolonged temperatures above seventy-five degrees (Fahrenheit) are intolerable for moose, so the recent hot summers in Maine have left them with little recourse except to find a cool place to lie down.

"When it gets too warm, moose typically seek shelter rather than foraging for nutritious foods needed to keep them healthy," the National Wildlife Federation (NWF) tells us. This leads to cows so underweight that they can't successfully bear a calf, reducing the moose population across the country. "Many biologists are concerned that they will have a difficult time adapting to climatic variability," NWF concludes.

So a visiting moose in a backyard may not be top on any homeowner's list of plans for their carefully cultivated garden, but at least it's a sign that the moose are seeking out food sources and eating a balanced diet. Take this as a good omen for a species that has faced more than its share of challenges for centuries.

What to Do—or Not Do—with a Visiting Moose

If you find yourself with a moose in your backyard, in a town park, on your street, or in front of you on a trail, there is no need to panic. In most seasons, moose will pay little attention to you as long as you maintain considerable distance. Here are the best ways to treat the situation:

- **Remember that this is a big, wild animal.** Do not approach the moose. If it perceives you as a threat, it is much more likely to get defensive and charge you. This is especially true of a cow moose and her young; if you see moose calves, make sure you are not standing between the mother and her children, and separate yourself from the situation as quickly as you can.
- **An approaching moose is not your friend.** If a moose sees you and starts to walk toward you, it either has some experience with people and expects a handout, or it wants to intimidate you into leaving before it charges. Either way, it's not coming over to be petted. If a moose calf approaches you, it may be curious about you—but this is not okay either, as it may lead its mother right to you. Back away and leave the moose behind.
- **Pay attention.** If you are in a wooded area where moose are known to live, keep an eye out for them. Moose may hear you coming and may decide to stand in place until you pass. In a forest, they may be entirely concealed by foliage and tree trunks.

- **Don't throw things**. Hurling rocks or snowballs at a moose can have exactly the opposite effect from what you mean to achieve. Rather than driving the animal away, this may be the proof it needs to see you as an attacker (which you are) and to take offensive action. If you have children with you, model the behavior you want them to follow: stay calm, treat the moose with respect, and leave it alone or move on.
- **Keep your dog under control**. Moose view dogs as enemies, so a barking or growling dog can prompt instant aggression from the moose, like charging or kicking the dog. If you see a moose while you're walking your dog, or if your dog is on a trail with you, give the moose plenty of room by choosing another route. Above all, keep your dog from chasing the moose.
- **If a moose blocks your path, wait**. If you're a safe distance away, waiting for the moose to leave will not seem threatening to it; it will probably ignore you completely. Moose can stand in one place for a very long time, however, so if it does not move on after half an hour or so, you may want to backtrack and take a different trail.
- **Watch for signs that the moose is annoyed**. When a moose gets upset, it lays its ears back, and the hairs on its rump stand on end. It may also lick its lips, but if you can see it doing this, you are much too close for comfort.
- **Talk softly and back away**. Let the moose know that you pose no threat to it. Raising your voice may have the opposite effect, so keep your voice low and back away slowly.
- **If it charges, run**. Unlike bears that will chase you if you run, moose usually lose interest once it's clear you're out of their area, so it's perfectly fine to run away. If you can't run, take cover behind something solid, like a big rock or a wide tree trunk. The moose may charge for a moment, but it will most likely be foiled and discouraged by the obstacle.
- **If it knocks you down, protect your head**. In the very unlikely event that the moose chases you and knocks you to the ground, curl up in a ball and put your arms around your head. Stay still

and wait. The moose may simply run past and on through the forest, or it may stop and kick you with all four hooves. Whatever happens, don't get up until you are certain it has moved a safe distance away, or it may turn around and attack again.

Moose encounters in the wild are unusual, but moose visiting yards, gardens, and orchards in the Northern states has become commonplace enough that blogs and an industry of deterrent products have sprung up around the subject. Maine, Michigan's UP, Utah, Idaho, Montana, Washington, and other north country areas offer us lots of advice for protecting plants, trees, and ourselves from these huge, voracious ungulates.

- **Build a fence.** If you really don't want moose or other animals to have access to your gardens, there's no better deterrent than a fence. Moose can leap over low fences, so plan on one at least eight feet tall; wire fencing works well, as hooved animals can't climb. If fencing your entire yard is beyond your budget, wire fences around specific trees will keep moose from attacking and devouring them—a particularly advisable precaution for fruit trees, which are very attractive to moose. Be sure to sink the fence into a trench in the ground several inches deep, to prevent the moose from trying to work its way underneath it. Some experts in deterring moose from backyards recommend using electric wire along the top of the fence, so moose attempting to lean over the top to get to a tree get a physical warning not to do so.
- **Protect your garden.** In addition to fencing, try placing cotton swabs saturated with wolf urine (available from Predator Pee or Deerbusters.com, for example) among your ornamental plants. Even in states with no wolf population, moose are likely to avoid anything that smells like a potentially lethal predator. Other moose repellants (like Pantyskyyd, a blood meal-based substance) are available with varying levels of stink; check online reviews before investing in something that smells so terrible it makes you want to avoid your own yard.

- **Try home remedies**. Some gardeners swear by placing pieces of Irish Spring soap around their vegetables and flowers to keep moose and other ungulates away. Another popular brew combines dish soap, water, and hot pepper (cayenne or chiles) in a spray bottle. Spritz this on your vegetable plants, ground-planted flowers, or fruiting vines to keep the animals away, and reapply regularly, especially after a rain. (If you use this for all kinds of animals, note that birds are not affected by the heat of the peppers, but mammals are.) Unused dryer sheets tied to tree branches are said to annoy moose enough to move them off. Moose may be repelled by these methods for several months but eventually learn to ignore them, so you may need to change up what you're doing from one season to the next.

- **If a moose is resting in your yard, let it be**. Moose may choose a shady spot to lie down, such as under a raised deck or up against a house or shed. If it's not bothering you, don't bother it; it will get up and walk away when it's ready. Repeated disturbances will only irritate it, which could lead to a confrontation. If you want to watch what it's doing, do so from inside your house or from across your yard.

- **Do not deliberately feed a moose**. It's one thing if a moose ambles into your vegetable garden and eats your cabbage, but it's quite another if you set out apples or carrots for it on your deck to draw it in closer for selfies. Feeding moose is illegal in some states and absolutely unadvisable across the country. A moose that become habituated to food provided by people will lose its healthy fear of humans, making its future behavior around people unpredictable, at best, and more aggressive, at worst, as it pursues handouts from any person it sees. The adage, "A fed moose is a dead moose" applies here; such a moose must be captured and destroyed before it hurts someone. The Alaska Department of Fish and Game does not mince words about this on its website: "By feeding a moose, people are more likely contributing to its death rather than its benefit."

Standing in front of the trap it had learned to ignore, this Eastern gray squirrel may or may not have lived in our neighbor's attic. © NIC MINETOR

Chapter 9

SQUIRRELS
CAPTAINS OF INDUSTRY

Bill and Joan, my next-door neighbors, awoke in the middle of the night to the *scritch, scritch, scritch* of something scrabbling its way through the inside of their bedroom walls. Instantly wide awake, Bill leapt out of bed, slid his feet into slippers, and made his way downstairs to investigate, grabbing a flashlight from the foyer table as he approached the basement. He guessed he had rodents afoot—rats or mice could have found their way into the wall of his city home in Rochester, New York, through some tiny slit between the cinder block walls and the basement ceiling. He found no such gaps, however; long-ago contractors had sealed up those openings before he and Joan bought the house. The critters, whatever they were, must have found a way in through the attic.

Panting as he climbed three flights of stairs from the basement to the top of the house, Bill immediately saw how animals had entered his home: moonlight streamed through a small but clearly gaping hole in the vent over the window at the east end of the attic. It didn't take long for him to find the chewed mess in a corner—a pile of newspapers, dead leaves, stringy plant remnants, and whatever else the small animals had gathered from the surrounding storage or carried in through the hole. He shone his light down into the gap between the outer and inner walls and found the culprits. An adult Eastern gray squirrel and at least three babies huddled together there, making the most of their comfortable shelter out of the wind and rain of early spring.

Bill knew he could do nothing right then to rid his home of the intruders, so he returned to bed and reported his find to Joan. She shuddered, but she agreed that any solution could wait until morning.

The next day, Bill visited the hardware store and bought a metal mesh trap with a one-way door, designed to trap small animals alive and hold them until they can be released somewhere else. He placed it in his backyard some distance from his home, baiting it with peanut butter, and sat down on his deck about fifty feet away to wait for an unsuspecting squirrel to go for the morsel inside.

It didn't take long. In minutes a squirrel approached the trap, giving it a wide berth at first, but coming closer when it spotted the lump of nut butter. Using the same stealthy method as a squirrel employs when it finds a new bird feeder hanging in my yard, it circled the trap, made feinting attempts to climb on it or enter it from the side, and finally worked its way to the trap's front door. From here it was an easy feat to slip under the door and reach the peanut butter. As the squirrel began enjoying the butter with gusto, the door slammed shut behind it, giving the animal barely enough leeway to pull its luxurious tail through the opening before locking it in.

From my own backyard, I watched the drama play out as the squirrel went into a panic, whirling around and around as it tried to find another exit. I heard Bill barely stifling a laugh from his seat on the deck.

"Trapping squirrels?" I asked. "You know, we live within walking distance of Wegmans if you need something for supper."

Bill snorted a one-note laugh. "I'm just gonna take him across town and let him out in a park," he said. "They're living in my attic, so I've gotta get rid of them."

"You think this is one of the ones in your attic?" I asked. "How can you tell?"

"I can't," he admitted. "But if I get rid of a bunch of squirrels in the neighborhood, I'm bound to get the ones from the attic."

I blinked. The logical flaws here had so many levels. "Have you thought about putting the trap in the attic? Then you can be sure you're getting the right squirrels."

He shrugged. "Or I could just catch them out here, so they don't go back in."

"But . . ."

I looked up at the third floor of his house and immediately noted the hole in the vent. "Is that how they're getting in?" I asked. "You could just repair the hole, which would keep any more from getting in, and then trap the ones still in there."

"And pay a contractor? Naw."

I gave Bill a congenial, neighborly smile. "Have fun, then," I said, and went off to my own pursuits.

Weeks later, the trap remained in the backyard, and I had seen it filled with frantic squirrels—as well as a raccoon, two cats, and a groundhog—a number of times on my way from the back door to our garage. When I next caught up with Bill, he had a rake in his hands, putting scattered spring detritus into paper bags.

"Still catching squirrels, huh?" I said.

Bill nodded once, breaking into a grin. "I've taken out more than thirty."

"Really?" I glanced up at his house and saw the hole in the vent, now edged with peeling paint and some evidence of gnawing. "You get the squirrels out of your house?"

"Not yet, but it's only a matter of time," said Bill.

I turned and faced him squarely. "Bill, you know I know a little about things like this, right?"

Joan had actually bought a copy of my book *Backyard Birding* from me over the summer, so it would be hard for him to deny that he knew I wrote books on nature topics. He lowered his jaw and looked over his glasses at me, the equivalent of a baby boomer eyeroll. "Yeah, I guess," he said.

"The squirrels will just keep multiplying. The more you remove from the area, the more the squirrels will sense a gap in the population, and the more squirrels they will produce. You're perpetuating a cycle that will continue *ad infinitum*." I stopped and slowed my speech, giving each word an extra punch. "You've become Part. Of. The. Problem."

Bill stood with his rake in his hands, his head cocked to one side, seeing me as if he had never really looked at me before. For a second, I thought he might give over. Then he shook his head. "That can't be right," he said.

I thought seriously about going into my home office and printing out a copy of the study conducted by zoologists at the University of Toronto, published in 2000 in the journal *Nature*. These scientists learned that when squirrel populations grew to the limits of what a given environment could support with food and shelter, the squirrels skipped their fall reproductive cycle. The arctic ground squirrels they studied waited until winter was over, during which many squirrels died—as they do every winter. In spring, the surviving animals found plenty of room in their environment for more squirrels, so they reproduced with vigor.

But trapping and relocating squirrels had become a mission in Bill's life, and facts would have no effect on that. So he continued to marvel at the steady, unchanging population of them in his yard despite his efforts to curtail them. He didn't fix the hole in the vent until a real estate agent insisted on it before he put the house up for sale a year later. Meanwhile, our neighborhood's healthy and varied squirrel population, with Eastern gray squirrels (*Sciurus carolinensis*), American red squirrels (*Tamiasciurus hudsonicus*), Eastern chipmunks (*Tamias striatus*), and groundhogs (*Marmota monax*, also known as woodchucks), still entertained us daily as they do today, plundering the nuts from our black walnut tree or robbing us of the variety of seeds at our many bird feeders.

The Truth About Squirrels

Squirrels are so much a part of human existence that this book could not be complete without acknowledging them. The planet contains an impressive 280 species of squirrels, from the most common tree, ground, and flying varieties to all kinds of squirrel-like creatures: chipmunks, prairie dogs, marmots, and groundhogs, to name a few. Just about any backyard in America may have its own population of habitat-appropriate squirrel or chipmunk, with a greater prevalence of ground squirrels and marmots in the West, and more tree squirrels, chipmunks, and groundhogs in the East.

Like many other small, furry animals, squirrels' lives center on three main activities: eating, procreating, and avoiding predators. They accomplish procreation without any help from the humans around them: we may notice a large nest of dead leaves and twigs stuffed between branches of a tall tree, looking as if a gust of wind would blow it right down; or squirrels entering and exiting a tree cavity excavated by an industrious woodpecker a year or so earlier. Here squirrels give birth to their offspring and keep them safe and warm until they can forage on their own, just ten to twelve weeks after their arrival.

The eating part of their lives, however, has much more to do with humans and the vast array of foods we offer to other creatures—birds in particular. All members of the squirrel family are primarily herbivores, feasting on a wide variety of seeds, nuts, buds, shoots, stems, roots, flower bulbs, mushrooms, and fruits, and augmenting their diet with the occasional insect or bird egg for added protein. This makes just about everything in the average city or suburban yard look pretty tasty to the resident squirrels.

These animals have their limits, though. Few squirrels take an interest in a vegetable garden—and if they do find something there that they can't resist, they will demonstrate that interest in broad daylight. If you go out to your tomato plants first thing in the morning and find that your ripe tomatoes have been sampled by an animal, the culprit is not a squirrel. Squirrels keep a diurnal schedule and do not roam your yard at night, so the thief must have been a nocturnal critter like a raccoon or opossum.

Squirrels will take advantage of every feeder and birdbath within their reach, always searching for the quickest and easiest path to sustenance. @ NIC MINETOR

Squirrels do enjoy flower bulbs, however, and may be found digging in gardens of tulips, gladioli, or many other plants that grow from fall-planted bulbs. If this is the case in your yard, consider changing what you plant in your garden: daffodils, for example, have bulbs that squirrels (and other animals) find quite inedible. Your local garden center can help you choose more plants that do not interest squirrels.

These inquisitive animals survive very well on their own in habitats ranging from forests to deserts, their days filled with foraging for food and caching much of it in preparation for less abundant seasons. You may stumble upon one of these caches in your own yard: a stash of sunflower seeds partially hidden under a brush pile, or a stack of walnuts still in their green husks buried under a shrub. These large hoards are called *larder caches*, something squirrels use to limit the amount of searching and digging they need to do in the dead of winter. Just as humans stock the fridge in anticipation of less hospitable or prosperous seasons, squirrels do this so they can see their food supply in one place and be sure that they have enough to keep them fed during a frozen January. Squirrels will guard each cache as fiercely as such a small animal can, fighting off other squirrels, groundhogs, birds, and chipmunks that might intend to plunder their hard-earned gains.

It's more likely, however, that you will not come across the many smaller caches a squirrel creates, as each may contain just one or two nuts or seeds—a behavior known as *scatter caching*. Burying small amounts of food in dozens of different places helps squirrels keep other animals from finding and plundering all of their food at once, ensuring that each squirrel will sustain its food supply throughout a long winter and into the early spring. So when you see a squirrel digging hole after hole in your grassy front yard, you're watching it act on its most basic survival instincts. This may not relieve your frustration at seeing these little animals messing up your carefully manicured lawn, but at least you have a better understanding of what they're up to. As the website of People for the Ethical Treatment of Animals (PETA) points out, "More than likely, your lawn will recover from the digging and benefit from aeration before you even have time to address any perceived problems."

A study conducted at the University of Exeter in England discovered that squirrels think strategically throughout the caching process, choosing a larder approach if they feel they are unobserved as they hoard their future food, and a scatter method if other thieving squirrels can see what they're doing. The chosen method also may differ from one species to the next—American red squirrels, for example, use larder caching to hold the cones they collect from a small stand of pine trees, keeping their food supply and its hiding place close together. Eastern gray squirrels and fox squirrels, on the other hand, prefer the scatter approach, which reflects the wider diversity of food sources they can tap. Their caches may range over a much wider area, with some close to dependable, consistently stocked bird feeders and some on the outskirts of a wooded area full of natural sources.

Here's where we learn that squirrels are even more remarkable than we imagined: a number of studies have revealed that no matter how many caches the squirrels create, they can remember where most of them are and return to them when they need them, even if it has been some time since they accessed them. One study determined that gray squirrels use visual cues such as landmarks to find their way back to their caches, relying on sight more than on their sense of smell as previously assumed (McQuade et al., 1986).

In a particularly remarkable study in 2017 at the University of Chester in England, researcher Pizza Ka Yee Chow raised squirrels in his lab and gave the adult squirrels a task: they had to figure out the right order in which to pull a series of levers to release hazelnuts they could see in a plexiglass puzzle box. Once they had this task mastered and the study concluded, the squirrels went back to their usual activities—until twenty-two months later, when Chow gave them a different puzzle box that could be solved using the same sequence of levers. The squirrels solved it quickly, clearly relying on their memory of the last time. "The solution . . . was the same as two years before," Chow told *Scientific American* in November 2023. "That's how we knew that they still remembered it."

This particular study gives us some clues about squirrels' determination to break into our bird feeders and devour their contents—and why they so often succeed. Squirrels can remember patterns and sequences, so they know what they tried before, what worked, and what didn't, making each subsequent approach to a tricky feeder a bit more successful than the previous one. They also can recall complex routes to reach the seed-filled treasure at the end, as the homegrown documentary *Daylight Robbery* did in 1988, when United Kingdom writer Jessica Holm tested the skills of wild squirrels in mastering an obstacle course to get to a full bird feeder. As the squirrel learned more and more skills, Holm lengthened the course and added new challenges, and the squirrel continued to execute one daring move after another to reach the prize. (The fascinating video and its sequel are still available on YouTube.)

This behavior is the hallmark of most backyard squirrel populations: the determination to find a way into whatever food source they target. They make a career out of finding and hoarding—yes, "squirreling away"—all the food they can collect throughout a season, so it's literally their job to dangle by their back feet from a wire to gather the sunflower seeds from a feeder, no matter how tricky it may be to reach it.

Luckily, an entire industry has developed to foil the squirrels and preserve our bird food for the birds we love. Squirrel-blocking feeders come in many shapes, sizes, levels of complexity, and price ranges, with varying levels of long-term success.

- **Weight-activated feeders** keep squirrels out by closing their feeding ports as soon as a squirrel sets foot on a perch. These remain wide open for birds that weigh an ounce or so, but they may have seesaw-like mechanisms that pull a shutter down over the feeding ports when a well-fed squirrel puts its full one-pound bulk on the lever or sliding cover. Most of these feeders are made of metal rather than wood or plastic (which squirrels will gnaw through before long), making them nearly indestructible. If you invest in one of these feeders, be sure to fasten it well to whatever supports it, as squirrels will learn to knock the feeder down, so its contents simply spill onto the ground.
- **Cage feeders** hold seed or suet in the middle with a large cage surrounding the food, with openings large enough for a sparrow, finch, or small woodpecker to pass through—but not large enough for squirrels. (You may see a diligent chipmunk make its way into the cage, as these much smaller, lighter animals can defeat most squirrel-proof feeders.) Again, these feeders need to be fastened securely

to a pole or hook, as squirrels will knock them to the ground and fiddle with the latch at the top of the cage until they spring it open. If placed too near a tree or shrub, these feeders can attract feral or domestic outdoor cats, who will sit in the nearby tree and wait to swat and kill any unsuspecting bird that enters the cage.

- **Trick feeders** are fairly expensive fun, with Droll Yankee's famed Yankee Flipper squirrel-proof feeder providing the gold standard. This feeder features a perch that encircles the bottom of the seed tube. The perch holds perfectly still for lightweight birds, but flies into rapid rotating action once a squirrel puts its weight on it. The perch spins until the squirrel finally lets go, which can take a hilarious few seconds. The fun ends fairly quickly, however, because once a squirrel has experienced that ride, it does not return—and squirrels observing it never go near it either. The Flipper's battery requires charging, though not frequently, as it spends most of its life in standby mode.

- **Baffles** cut off a squirrel's access to a feeder by blocking its route up a pole. Birding stores are full of all kinds of baffles: bell-shaped baffles that hang below the feeder, preventing the squirrel from reaching the feeder by creating too wide an obstacle for it to get around; large domes that hang above the feeder, keeping the squirrel from accessing it from the top; and flat baffles that block the squirrel from walking the tightrope (a clothesline, for example) to reach a feeder hanging some distance down the line. All of these slow down the squirrels with the potential to continue to work indefinitely with no further action on your part. Choose baffles made of strong metal, Plexiglas®, or the hardest plastics to make them impervious to squirrel teeth.

- **Chemicals**. While the squirrel-busting industry markets a number of liquid, pump spray, and powdered substances as deterrents, the only one we have found to be effective consistently is capsaicin, a natural chemical that gives the heat to spicy peppers. Birds do not experience the heat and actually seem to enjoy the flavor, but squirrels feel the burning and detest it, dropping from the feeder to the ground and desperately trying to wipe themselves clean of it. Once burned, twice shy—we find that the squirrels do not return to that feeder for months, even if we stop spraying the seed or nuts with capsaicin. In our experience, the bottled liquid is stronger and more effective than the seeds that are pretreated with pepper spray.

Other chemical deterrents that we have found less successful include mesh bags of mothballs; an ammonia-soaked rag placed near feeders; bottled coyote urine, which is enough to scare off many rodents and other small furry animals; and organic squirrel repellants sold by garden centers and some birding specialty stores.

When Your Home Is Their Home

Squirrels clearing out bird feeders is a familiar story to homeowners in any community with wildlife, but even the most fascinating squirrel acrobatics lose their charm when the little devils chew their way into a resident's home. Suddenly they go from being clever thieves to vermin, their presence an infestation of the house that creates a shivery sense of the unclean, skittering their way up and down the walls.

Some unnerved homeowners want to move directly to poisoning and killing squirrels, chipmunks, and groundhogs instead of finding a way to discourage them from their homes and property. Undoubtedly, poisoning squirrels will prevent them from returning to your home, so it may seem like the swift and even humane solution—certainly more so than kill traps that break their necks on contact. Using deadly chemicals has far-reaching consequences, however, including a serious one that reaches into the already precarious world of birds of prey.

Owls, hawks, foxes, coyotes, and even some outdoor cats (what animal control experts call "nontarget animals") all eat rodents. Rats, mice, moles, voles, shrews, lemmings, chipmunks, and squirrels all make up the largest part of raptors' diets—and a lethargic squirrel just beginning to feel the effects of a rodenticide makes a very easy target. The difficulty, however, develops after the bird or animal digests the poisoned rodent. This noxious poison quickly works its way into the system of the hungry bird or cat, becoming a "second generation anticoagulant rodenticide," or

SGAR. "What's unique about [SGARs] is that rodents frequently eat more than a single dose of them, and the effects of that dose are often delayed for a few days," the California Department of Pesticide Regulation explains on the Golden Gate Bird Alliance website. "Meanwhile, the rodents may continue to eat more poison, resulting in a super-lethal dose that builds up in their tissues. When predators such as hawks or foxes eat these weakened or dead rodents, the dose may also be deadly to the predators. Incident reports conclude that SGAR products pose significant risks to non-target wildlife and that these risks are greater than those posed by other rodenticide active ingredients."

SGARs prevent blood from clotting, so the affected animal bleeds internally over the course of several days after ingesting the poison. As the substance does its work, the animal's condition proceeds to organ failure, paralysis, and eventually death. There is nothing humane about this poisoning process—especially if the dying creature is the raptor or mammal that simply ate the poisoned rodent. SGARs are now banned in California as well as in other states—but they are still in use in much of the United States. So if you choose to use poisons to kill squirrels, you may be killing a significant segment of your backyard ecosystem.

Here in upstate New York, people like my neighbors Bill and Joan can choose a pest control company that does not use poisons to remove animals from homes. The method involves finding and repairing the damage the squirrels did to enter the home, and replacing chewed fascia board, shingles, or other materials with screening to keep more animals from using the route as an entry point. Pest control professionals install one-way doors in the remaining openings to allow the squirrels an easy exit with no return entry, waiting until all the squirrels have left the building before beginning final repairs and cleanup of their indoor nesting area. They finish by sealing and repairing the entry point, ending the squirrel problem without killing anything.

Canada geese have made a startling comeback across the continent after being hunted nearly to extinction. © NIC MINETOR

Chapter 10

NO GOOSE, NO GOOSE...
GIANT GOOSE!
THE CANADA GOOSE RETURNS

Cars came to a complete stop in the middle of a busy intersection at Marketplace Mall in Henrietta, New York, on a balmy May morning, one of hundreds of identical mornings at this well-landscaped shopping center. Sparkling ponds provide welcome eye relief between the expanses of gray asphalt parking lots and the carefully mowed grass lawns, an effort by the owners to make this sprawling complex seem a bit less of an intrusion on the farmland it replaced back in 1982.

This is upstate New York, however, and wherever mowed grass dominates the earth, large, waddling, mostly brown birds feel entirely free to stop traffic. These boldly striking waterfowl strut with their grayish chests erect and long necks regally raised, their black faces interrupted with a wide white chinstrap that wraps around to the top of their head. Today they have taken control of the intersection, stopping a long line of cars as two adults march across eight lanes of traffic. To the delight of drivers at the head of the line, a row of fuzzy yellow goslings follows them, their legs scurrying in triple time as they rush to keep up with their parents. I count eight, nine... twelve tiny chicks leaving the relative comfort of the grass to step on pavement for the first time.

"Make way for goslings!" I can't help but say aloud to no one in particular, though the woman driving the car next to mine smiles and nods to me. She's about my age; we may have watched the same animated telling of Robert McCloskey's classic book *Make Way for Ducklings* on children's television fifty-five years earlier. Baby birds are one of the great equalizers, I think as I wait for them to pass—one thing virtually all people have in common is our immediate, instinctive response to protect any kind of baby animal, a need to come to the mother's aid and to make sure that the little ones stay out of harm's way.

The goslings reach the other side safely and begin to hop into the pond there. Despite the time lost to this little parade, some cars still hesitate as their drivers and passengers watch each tiny chick launch itself into the water, where a parent is already swimming and clucking to her brood. The last adult Canada goose (*Branta canadensis*) steps off the road and runs forward a bit to shepherd the stragglers off the bank, standing by until the last one splashes in and paddles off. More geese will grow into adults with the skills they need to survive in this environment—skills like nibbling grass, staying out of the way of oncoming traffic, and hissing at passers-by to let them know who's in charge here.

Oh, and defecating on every sidewalk. Let's not forget that.

If it weren't for this penchant for leaving their waste just where people want to walk, the large flocks of Canada geese (that's the correct name, *not* Canadian geese) might be a welcome sight to suburban and urban residents all over the country. These geese, after all, have long been harbingers of the coming spring and the waning fall, lifting off of open fields and flying in their regimented V formations across the autumn sky. Their loud, unmistakable honking as they fly has become one of the iconic sounds of the changing seasons. For more southerly communities that host millions of geese over the winter, the bountiful hunting season provides a draw for tourism, and a considerable supply of meat for those who enjoy goose's rich, lean taste and similarity to beef or venison. Cooking a wild goose requires a long, slow roast to restore tenderness to a bird toughened by daily flying, but connoisseurs claim that it makes a tasty stew or casserole.

To their own detriment, however, geese don't pay any attention to the manmade accoutrements in their environment, seeing each of them as just another place to leave their slimy mark. Combine this with the birds' fierce defense of their goslings until the little ones have reached full-grown adulthood, displaying with wings raised and hissing cries when people get too close, and the birds leave some people less grateful than we might wish to have Canada geese in their neighborhoods.

Perhaps these folks would feel differently if they had a better understanding of how close we came to losing Canada geese altogether. As recently as the early 1900s, hunters and their families knew well how delicious wild goose tasted on the Sunday dinner table. Like the passenger pigeon and the Carolina parakeet, Canada geese flew together in flocks of thousands of birds, making themselves very easy targets for hunting parties and their many guns. With no regulations on how many geese they could kill in any given season, parties of hunters feeding their families or their communities shot down hundreds in a single day, faster than these birds could reproduce.

At the same time, farms, cities, and towns sprang up where open fields had accommodated huge flocks of geese throughout the spring, summer, and fall months. Migrating geese returning from their winter grounds on the Southern states' agricultural fields found few such lands in the North, so they could not stop to rest before pushing on. Many geese could not finish the journey, especially those that crossed the Great Lakes into Canada or followed the Atlantic Flyway into New England. Soon, Canada geese vanished from the Northern skies—and with no reproduction in the North, the Southern states lost their flocks as well.

One more human practice contributed to the death knell for these mighty geese: egg collecting. Canada geese lay clutches of ten or more eggs, one of the secrets of their ability to maintain their flock sizes even with all of the challenges they have faced. If these eggs are not left to incubate and hatch, however, they obviously cannot produce young. When families needing food began sending out their farmhands and children to collect the goose eggs shortly after they were laid, the geese were left with no broods. By the 1910s, many flocks of Canada geese had simply stopped reproducing, and their numbers fell as hunters took the remaining birds. It seemed that the world's largest goose would go the way of the dodo.

Then a series of extraordinary things happened. First, the federal government passed the Migratory Bird Treaty Act of 1918, making it unlawful in the United States to hunt, kill, purchase, or sell any native bird that migrates through North America except during specific seasons;

even then, hunters had to secure a special permit and obey strict limits on how many geese they could take per day. The protection even extended to the birds' nests and eggs, giving the Canada goose and many other birds a fighting chance of restoring their own numbers naturally.

Still, some of the eleven subspecies of Canada goose had declined to a point at which the scientific community presumed they were gone forever. In particular, a subspecies known as the "giant" Canada goose, *B. canadensis maxima*, seemed to have disappeared completely. Giant Canada geese weigh three or four more pounds than "normal" Canada geese, often reaching sixteen pounds instead of *B. canadensis*'s twelve pounds. One account even suggests that these geese could top out at twenty pounds, making them unmistakable in a mixed flock of Canadas.

"They were all big geese," wrote William B. Mershon, a businessman from Saginaw, Michigan, who also served as a state forester and a member of the American Ornithologists' Union (AOU). "We knew them as such. While having the general markings of the Canada goose, such as the black head and neck with white throat and cheeks, they were lighter colored, of a blue-ashy general appearance, and the bodies were shaped differently—long ovals instead of round, chunky ovals ... They usually flew low, not so high in the air as the other geese. If we had a shot at the bunch of them, the geese we got were all big ones." This observation gives us a clue to why these largest of the Canada geese were so much more vulnerable to the hunters' bullets, apparently bringing their existence to an untimely end.

What ornithologists of the time did not know, however, was that in 1924, Dr. Charles Mayo—one of the founders of the Mayo Clinic in Rochester, Minnesota—had acquired fifteen of these giant Canada geese from North Dakota. This made the giants a particularly majestic sight, and just the kind of creature Mayo desired to grace the lakes on his Mayowood estate.

Taking full advantage of an artificial feeding program Mayo initiated, the giant geese thrived in Rochester, attracting migrating geese to join them on Mayo's lake. By 1939, hundreds of giant Canada geese had gathered there. Mayo discontinued the feeding program, but the geese stayed anyway, foregoing migration to overwinter here on water warmed by a newly constructed power plant.

Nearby residents knew of this flock but were oblivious of its significance, so they paid it scant attention. Word of it had not even reached the AOU by the publication of its annual checklist in 1957. "Now believed to be extinct," the document noted, a somber footnote on the life of a remarkable bird.

In 1962, however, the fate of the *maxima* subspecies changed dramatically when Harold Hanson, a research biologist with the Illinois Natural History Survey, was invited to join members of the Minnesota Department of Conservation to band, weigh, and measure the Canada goose flock on Rochester's Silver Lake. As they began working with this flock, they reached a perplexing conclusion: The scales they were using had to be off by several pounds. How could these Canada geese weigh as much as the giant geese that had long since been declared extinct? One of the men paid a visit to the local grocery store and purchased a ten-pound bag of flour and a five-pound bag of sugar, weighing them at the store to be certain the measurements were correct. He used these to test the laboratory scales, and to all of the conservationists' amazement, the scales were indeed accurate. These geese had to be *B. canadensis maxima*!

The discovery led Hanson to publish a book about the entire life cycle and reemergence of giant Canada geese, a subspecies he noted was not extinct at all, but "still moderately abundant in parts of the prairie provinces of Canada and in some areas of the United States." This work led to continent-wide operations to restore the subspecies to its original habitats, a process that benefitted all Canada geese, increasing their numbers to the abundance they enjoy today.

Minnesota's giant geese were just one of the successful repatriations of these birds in America. Back in the early 1900s, New York State's wild goose population had dwindled to just a few nesting pairs, descended from birds owned by some of the state's wealthy residents who had their own collections of waterfowl. These few released birds formed new flocks in the Lower Hudson River Valley and began to move northward into less populated, more open areas of the state. In the 1950s and 1960s, as Minnesota's work to understand its giant goose populations got underway, New York's State Conservation Department released geese that had been raised on game farms, transporting them to wildlife management areas north and west of Albany. These geese adapted well to their new surroundings and propagated with great vigor. Today more than two hundred thousand Canada geese live across New York, making the most of farm fields, golf courses, open lawns, wetlands, and other areas that provide plenty of food and space for these birds.

The restoration of goose numbers across the continent has been hailed as one of the great success stories of the conservation movement, but some see it as a case of species recovery gone very wrong. The New York State Department of Environmental Conservation (NYSDEC), for example, notes on a page of its website titled "Nuisance Canada Geese" that these birds are "a valuable natural resource that provides recreation and enjoyment to bird watchers, hunters, and the general public." With this established, it goes on to point out that many geese have discovered that the state offers them a year-round food supply of open, grassy lawns; lakes and ponds that don't freeze over in late fall as they once did; and farm fields that provide abundant grain and vegetables left on the ground after the harvest.

Any lake, pond, or gentle river may provide a place for Canada geese to nest and produce litters of goslings, even in someone's backyard. © NIC MINETOR

This holds true in many communities throughout the country. Canada geese have made year-round homes where they once just migrated through, stopping briefly on their way to nesting grounds much farther north and wintering areas farther south. This proliferation of year-round geese infuriates human residents who face the damage and the mess these birds make: overgrazed lawns and slippery droppings on sidewalks, playgrounds, and paved fitness pathways. One goose can produce up to two pounds of fecal matter every day, and while it may seem like all of this covers the sidewalks throughout suburban neighborhoods, much of it actually ends up in ponds and lakes. This can spread waterborne diseases, promote algal blooms, and generally mess up the water supply.

Large goose flocks moving from one field to another also raise safety concerns along roadsides and in the air. The possibility of a serious accident involving a goose became crystal clear on January 15, 2009, when the collision of US Airways Flight 1549 with a flock of Canada geese shortly after the flight's takeoff from New York's LaGuardia Airport disabled both of the plane's engines. Captain Chesley "Sully" Sullenberger thought quickly and landed the Airbus A320 in the middle of the Hudson River, a feat that became known as the Miracle on the Hudson. Only five of the 150 passengers and one flight attendant suffered injuries, but no one died—except the geese, which were transformed by the force of the engines into a substance aviators aptly call "snarge."

So why are geese still protected under the Migratory Bird Treaty Act, now that they have obviously recovered spectacularly? This treaty is international law, covering the entire world—so birds considered a nuisance in North America may be rare or endangered in other countries. Birds have no concept of borders or nations, so they migrate to wherever they find habitat that provides what they need, many of them following instinctive routes that their ancestors have traversed for thousands of years. Each country is bound by the treaty to protect them, no matter how many of these birds each may host. Changing such a treaty requires all of the member countries to agree, which would be unlikely even in a perfect world.

There is always the chance, however, that changing the protections for a specific bird species could mean that that bird slips back into the endangered zone in the future. Canada geese are not the only birds that are considered a problem for humans, and some of the others—cormorants, grackles, blackbirds, pelicans, starlings, cowbirds, house sparrows, and more—have had their numbers drop to critical lows in some countries, while their populations soar in others. All birds continue to face dangers from pesticides, window strikes, and house cats, so keeping the treaty in place for all birds allows wildlife conservation organizations to take action to ban or modify some of these threats, protecting every bird in every country.

How to Manage Geese on and Around Your Property

State wildlife departments have tried many different ways to keep geese from settling in areas where they are most undesirable, like golf courses and developed parks—but there is no single, surefire way to eliminate the geese from any area they find attractive.

Geese have no fear of humans, so the kinds of hazing techniques used to scare away deer, turkeys, or other animals may be useful in the short term but will not trouble the birds in the long run. Banging pots and pans together, motion-activated sprinklers, or noisemaking systems may deter geese when they are new to an area, but they will have little or no effect on geese long established on a particular field or lawn.

Here are some goose management methods suggested by experts across the country.

- **Don't feed the geese.** Geese find all the nutrition they need in lawns and fields with low grass or crop stubble, so they don't require corn, bird food, or bread from you to survive. Like most animals, they will take the path of least resistance to a meal, so if you feed them, they will gather around you and stay there until the food is gone. This is unhealthy, however: Geese rubbing up against one another to gobble up your cracked corn can spread illnesses among the flock.

    Let's be clear: Bread, crackers, and popcorn are all bad for birds, as their digestive systems are not made to break down processed foods like these. The birds stay full of bread longer, so

they stop seeking the nutrition they need from their normal diet. Feeding them bread just leads to malnourished birds.
- **Let your lawn go.** Geese love a mowed lawn, so stop mowing. If you live where you can let your yard become a place for tall wildflowers, you can end your goose problem by rototilling the grass and scattering native wildflower seeds (or if you live in an arid region, cover the ground with gravel). A natural meadow full of tall plants and shrubs holds no interest for birds like Canada geese, and a xeriscape of pebbles and ground covers will turn them off completely. This also provides the added benefit of reducing your time behind the mower. Think of the free time you'll have when you stop manicuring your lawn.
- **Hang things over their heads.** "Geese are normally reluctant to linger beneath an object hovering overhead," notes NYSDEC, so attaching certain kinds of mylar tape or balloons to poles can scare them away. NYSDEC recommends balloons imprinted with a large eye spot, making them look weird and threatening to the birds. (Some garden centers and DIY hardware stores sell these; look for balloons thirty inches or more in diameter, filled with helium.)
- **Create a grid over your pond.** If geese regularly rest on a pond on your property, you can deter them from doing so by installing a grid of crisscrossed wires over the water. NYSDEC recommends single strands of #14 wire strung ten to fifteen feet apart, secured by posts that keep the wire at least eighteen inches above the surface of the water. This will prevent the birds from landing on your pond, which will make the adjacent lawn or fields of stubble less interesting to them. Add bright-colored flags or streamers to the grid to keep the birds from attempting to land and becoming entangled in the wires. (This deterrent may not be a good solution if you or your family use the pond for swimming, fishing, playing with remote-controlled boats, or any other recreational activities.)
- **Add a pond fence.** If geese land on your property's water feature and walk up onto land from the water, a fence on land close to the

edge of the water will keep them from doing so. This works well in summer, when geese molt and lose the ability to fly for weeks at a time. A tight, dense hedge of shrubs can be more attractive than a fence and just as effective.

- **Make a joyful noise.** Projectile firecrackers, bird-bangers, and other loud pyrotechnics can annoy birds enough to convince them to leave you alone. Consult with a fireworks distributor who has experience in discouraging birds to be sure you are buying something that will be effective in this case. (Pyrotechnics may not be an appropriate solution for residential areas, and they may be illegal if fireworks are prohibited in your state.)

- **Get a dog that chases geese.** Dogs with herding instincts, like border collies, can be trained to chase geese off your property. This is a labor-intensive solution, as your dog needs to be trained to do this, and then you and the dog must chase the geese several times a day for weeks before the message gets through to the birds. Once you have succeeded in harassing the geese enough for them to vacate, you will need to continue to patrol your property regularly to keep them from coming back. If you can commit to this level of policing your geese, NYSDEC says that this method is one of the most effective ways to get rid of the birds.

- **Sour the grass.** One goose repellent has been approved by the US Environmental Protection Agency for use on lawns. It's called ReJeXiT®, and it makes the grass taste bad to geese. To be effective in getting geese to leave your property, you will need to spread the repellent on your entire lawn—which will get expensive if you have a lot of lawn to cover, as it costs about $145 per acre (2025 pricing). Your local Department of Natural Resources (DNR) may require you to have a permit to apply it; contact them for more information.

- **Discourage or disrupt nesting.** Geese come back to the same area to nest every year, so you can predict their behavior and be ready to act. Harassing the geese with noisemakers or balloons before they choose a nesting site may discourage them from doing so right where you're carrying on this activity. They may go just beyond your reach to nest, however, so this is only a partial solution.

A more effective method of discouraging nesting involves more direct action: you can sign up at the US Fish and Wildlife Service's Resident Canada Goose Nest and Egg Registration Site at https://www.fws.gov/eRCGR/ to receive permission to destroy the goose's eggs and nest. This may be the best way to rid your property of a goose defending a nest near a doorway or on a playground, or to simply discourage the goose from continuing to nest on your property. Several methods are suggested at the US Department of Agriculture's Animal and Plant Health Inspection Service's website at https://www.aphis.usda.gov/sites/default/files/canada_goose.pdf, including treatment of the eggs with corn oil, a nearly foolproof way to keep them from hatching. (Visit the USDA website to learn the best methods for marking, oiling, shaking, and/or puncturing the eggs, as there are right and wrong ways to do this.) It's important to leave the eggs in place and simply render them unviable, so the goose will continue to incubate them rather than laying more. Once the eggs do not hatch, the goose will eventually give up and will not attempt another brood that year.

Preventing the geese from producing young helps reduce the number of birds on your property for the long term, but it helps in other ways as well: geese are less likely to return to a nest site where they did not produce young successfully, so this encourages them to move on. If you can engage your neighbors with geese or even your entire town in this effort, you may be able to curtail the growth of your resident goose population. (It may take five years or more for tangible benefits to become clear.)

- **Catch and release.** You may be able to acquire a permit from your state to catch the geese on your property and move them to another area. Moving geese off your property and onto someone else's land is not a solution; it's just a redistribution of the problem. You will need to work with your local department of natural resources to determine a place to take the geese that does not make them someone else's headache.

Few animals have adapted as brilliantly to living among humans as the raccoon, the clever rapscallion of urban nights. @ 6381380, ISTOCK

Chapter 11

THIEVES IN THE NIGHT
RACCOONS

On a snowy night somewhere in suburban North America, a man snaps on his porch light and begins to open his back door. (I am not sharing the man's name or location because I did not gain permission to do so. For the record, I tried several methods to contact him for an interview, but I received no response.) The door opens outward, but it's blocked by some kind of plushy obstruction—after a few attempts, he manages to get the door open far enough to shove his hand through the gap. In that hand he holds a bowl, and he quickly dumps it over as the wiggling mass against the door moves back and away. Now we can see the soft, furry blockade: a cluster of unusually plump adult raccoons (*Procyon lotor*) that scrabble across the deck floor, chasing the grapes the man rained down on them.

With the raccoons thus distracted, the man picks up a plastic tub large enough to hold a ten-pound turkey—but it's filled to the top with hot dogs. He tells the camera that he purchased twenty pounds of these today just for his raccoon pals. As the animals—we can count about twenty-five of them, but it looks like there may be even more off-camera—make way for their host, he steps outside onto the frost-covered deck and crosses to a bench, where he sits down as the raccoons swarm him, vying to be the first to receive their meal. The man begins passing out dozens of hot dogs until every raccoon has at least one, and as the furry creatures start lining up for seconds, he has the opportunity to pet some of them, including one that has perched on the back of the bench above his shoulder and now leans over him, ready to snatch anything the others don't get.

These raccoons clearly know the man well; they know what to expect and what role they need to play to get second and third helpings. In a few minutes, all the hot dogs are gone. The feast isn't over, though: the raccoons stay for a second course of crackers and what the man describes as "dry" food, presumably something made for dogs; they continue to munch contentedly, snuffling and grunting across the porch floor as their host returns to his warm kitchen.

It's clear that this happens nightly, and that it's gone on for quite some time. The raccoons are well conditioned to wait on the porch, take turns retrieving food, and even tolerate some physical contact with their benefactor. What's not clear in this interchange is whether these animals still scavenge in the surrounding woods or the neighborhood for their own food, or if this man's uncommon generosity has become the entire source of their livelihood. As he goes back inside, the man tells viewers that, in his experience, as the weather gets colder in the coming days and weeks, the raccoons will move off to wherever they spend the winter and he won't see them again until they bring their young to his porch the following year.

This YouTube video has generated an astonishing forty-two million views, with hundreds of thousands of kind and encouraging comments from amused viewers. Indeed, it's great fun to watch these remarkably polite and orderly animals gather around this man, extend their eerily human-looking hands to receive their hot dog, and move away to a clear patch of porch to eat it, allowing others to step up for their share. Learning manners acceptable to humans may never have been on their list of aspirations, but they clearly know how they have to behave to continue to enjoy this easy, perpetual food source.

Here Comes the Buzzkill

As undisputably cute and overwhelmingly popular as this video is, it sets off a whole set of alarm bells for wildlife professionals. How many of the millions of viewers will be encouraged by this to start feeding wild animals off of their own back porches? A Facebook page devoted to raccoon feeding has sprung up, with around 9,600 members—some of whom post videos and photos of raccoons and opossums feeding out of pet food bowls on their decks. The man with the hot dogs became an internet sensation during the pandemic lockdown, when people around the world barricaded themselves in their homes, looked out a window, and saw things outside that fascinated them. Here was a man who developed a relationship with animals that clearly filled his time in daylight as well as during their nocturnal feeding time: he shopped for various foods for his raccoons, planned how he would provide the food to the animals, sat with them and took in the phenomenon he had created, and had more than two dozen furry friends. These choices appealed to lonely folks throughout North America, and perhaps even beyond.

But what will happen to these raccoons if the man moves away from the home they rely on for their daily feast? Do they still know how to scavenge for their own meals, or has this new feeding pattern erased that skill from their collective memory? What will the next homeowners have to contend with if more than twenty-five raccoons continue to show up at their back door night after night?

On the other hand, is handing out hot dogs and cookies to these animals any different from leaving our trash cans unsealed and unattended, so raccoons learn to ransack these containers to find their own spoils?

Wildlife biologist Caitlin May of the US Fish and Wildlife Service clears up these ambiguities in an article she wrote for the agency's website. She explains that every animal has a specialized diet of foods found in the wild that provide the nutrition each specific species requires. Whether these animals eat insects, plants, or other animals depends on these needs—but an animal accustomed to eating mice, rats, and voles will not find the same nutrition in an ultra-processed, cooked food like a hot dog. "Food designed for humans or domestic pets is meant to meet the nutritional needs of the species it is designed for," she wrote. "If a wild animal were to eat food like this, they might not get the nutrients they need in order to survive." The result is malnourishment, even when the human-fed animal appears quite chubby from its diet of high-fat human food. Malnourished animals grow weak and eventually fall prey to the next animal on the food chain, or they develop health issues that would be unheard of if they stuck to their normal food sources.

This is equally true for pet food, which contains nutrients that domestic animals require because they do not hunt for their own food in the wild. Just as your backyard birds do not require fortified seed blends made for caged parakeets, raccoons should not eat dog or cat food bolstered with vitamins. Wild animals acquire all of these nutrients from the varied diet they enjoy in the wild. "Pet food is just as harmful to wildlife as human food," May underscores.

Human food that animals find accidentally—in trash cans or dumps, for example—is no better than whatever food they have been conditioned to take from a person's hand. Generally, animals have no way to understand packaging materials, so they may or may not rip through the package to eat what is inside—they may simply eat whatever is between them and the food they can smell through the package. When raccoons eat plastic wrap, foil, or paper, these undigestible materials can make them sick.

As if all of these things were not reason enough to avoid feeding raccoons, other dangers can happen to the animals and to the people feeding

Raccoons find all kinds of places in the city to call their own, including storm drains, where they can raise their young completely out of sight of predators and human neighbors. © CHRISTINA RADCLIFFE, ISTOCK

them. Animals crowding together to receive food from people can pass illnesses between them, like distemper, hantavirus, and rabies. Raccoons often will huddle together in family groups, but not in whole colonies like the one that has formed around the hot dog feeding ritual. Some of these diseases can pass to the humans interacting with the animals.

No matter how endearing the raccoon feeding video may be, it's not a healthy practice for the man or his raccoons. It's also not necessary, as raccoons are among the cleverest and most resourceful animals in North America, perfectly capable of making their own way in a world that grows more complicated for them every year.

THE KINGS OF URBAN WILDLIFE

Even before they had evolved into the animals we know today, raccoon ancestors crossing the Bering Strait (or the land bridge that preceded it) from Russia to Alaska six million years ago had the ability to make the most of their surroundings. Fossil records tell us that procyonids—small tree-climbing mammals—moved southward to tropical areas, living in these warmer climates until about 2.5 million years ago, fairly recently in geologic time. As evolution progressed, crab-eating raccoons remained in warmer climates along the oceans, while others chose a different evolutionary path and made their way north, settling in the Great Plains. Eventually, they congregated along large rivers and in the Southeastern states, choosing areas with high concentrations of tall trees for bearing and raising young out of the reach of predators. Here America's Indigenous people encountered them, and the Algonquian Powhatan people named them *arakunem*, or "he who rubs, scrubs, and scratches with his hands"—a reference to their peculiar behavior of washing their food in a stream or pond before eating it.

The raccoon's luxurious fur and flamboyantly striped tail quickly attracted the interest of European fur traders. Indigenous people already traded furs among the tribes for other goods they needed, so they offered this same arrangement to the European colonists, introducing these newcomers to beaver, marten, mink, and otter as well as raccoon, and receiving knives, kettles, blankets, and other goods in return. As the Hudson's Bay Company and other commercial enterprises began to grow in

North America in the late 1600s, fur trapping and trading became one of the primary industries of interest. Fur coats and hats became all the rage in Western Europe's fashionable circles. Raccoon fur made good hats, noted naturalist John Lawson in 1714, and the animals' skin made "fine Women's shooes [sic]." While most fur commerce centered on the beaver—enough so that the French and Indian Wars also have the nickname "Beaver Wars," as the conflict centered on control of the beaver pelt trade—raccoons also fell to trappers in massive numbers. Coats made from raccoon pelts rose in popularity with the European lower classes who could not afford the higher-priced beaver, while the "coonskin cap" worn by many rebel patriots and frontiersmen in the mid-1700s made raccoons very much in demand on both sides of the Atlantic Ocean.

By this time, raccoons had moved up from the Southeast and established themselves in the Great Lakes region, where they became more accessible to fur trappers and traders. This period of prosperity for fur traders came to an end in the late 1700s, however, as settlers cleared so much land for settlement that fur-bearing animals retreated into the few remaining forests and did not bear as many young. Over-trapping also reduced the number of adults of reproductive age of many fur-bearing animals, bringing the entire industry to a fairly sudden end. European haberdashers and milliners stopped lining hats and coats with animal skins, switching easily to silk and rejecting the high prices of the trapping industry. By the mid-1800s, the fur trade in America had virtually disappeared, and the last fans of raccoon coats—Jewish people in Russia and Poland—lost access to their preferred furs in 1847, when the Russian czar banned the use of raccoon throughout the country.

The ring-tailed bandits faced an additional brief period as targets, however, when raccoon coats became all the rage on college campuses in the 1920s. The rise in automobile travel made heavy coats necessary, as most cars were wide open with no roof, so driving in winter became a frigid affair at best. The fad lasted through the end of the decade, but with the onset of the Great Depression in late 1929 and the development of enclosed automobiles with hard tops and heaters, raccoon coats went into mothball-equipped closets and were soon forgotten. They had another brief hurrah in the 1950s as college students discovered these garments in the backs of their parents' wardrobes, but all fur coats soon lost their popularity in the mid- to late twentieth century, as activists took up the cause of abused animals raised in deplorable conditions on fur farms. Today some states have gone so far as to ban fur farming altogether, sounding the death knell for the already shrunken industry.

Raccoons Learn to Assimilate

With their numbers growing as the fur trade fizzled in the mid-1800s, clever raccoons discovered the advantages of living near humans in urban areas sooner than most other mammals did. They made the most of the lack of sanitation services in large cities in the nineteenth century, plundering the trash they found dumped in the streets. Even as city infrastructure improved in the 1880s, however, and the sanitation industry became well established, raccoons learned many tricks that made cohabiting with human beings the easiest route to a full belly. They munched on vegetable gardens, found dumpsters that never closed, and even romanced some humans into feeding them by hand. It pays to be clever, and no urban animal is cleverer than the so-called trash panda.

When their natural predators like wolves and mountain lions fell to bounty hunters in the 1940s (see chapter 2 for more on this), raccoon populations expanded rapidly. Today raccoons are in all of the contiguous forty-eight states and in most of Canada, even establishing new habitats in the Rocky Mountains. Every city, suburb, and rural area in the continental United States hosts a healthy population of raccoons. They transferred their need to live in trees to attics, sheds, barn lofts, and garages, and eventually to low shelters like crawlspaces, basements, storm drains, and under porches, finding safety there from dogs and hawks. Soon they realized that even the few remaining wild predators did not venture into congested, human-populated areas, so cities became safe havens for raising baby raccoons. They have found their way into whatever receptacle or structure they have chosen, squeezing through openings no larger than three inches across. The answer to foiling raccoons, then, becomes simple: don't give them that opening.

"Raccoons don't know that our trash cans, vegetable gardens, bird feeders, and chimneys aren't for them—they're just trying to survive," the

Humane Society of the United States tells us on its website. There's no malicious intent in their actions; their trespass is proof of their adaptation to the new kinds of shelter that show up in well-established raccoon habitat.

Marked by nature with a black stripe across the face that resembles a mask, raccoons can't help but be seen as somehow sinister, skulking and rustling about in the dark like thieves and plundering anything they can reach with their hand-like paws. If you watch them emerge from their dens at dusk, however, as my husband and I have had the opportunity to do for decades from the comfort of our front porch, you see the careful observation that goes into every move they make. Raccoons may disrupt your yard and make a mess of your trash, but none of this behavior is personal—they are simply willing to do whatever is necessary to feed themselves and their families, and to provide shelter to allow their offspring to thrive and grow.

How to Manage Your Relationship with Raccoons

Your raccoon neighbors are here to stay, but you're not obligated to serve them dinner every night or provide them with shelter. Here are some tips from experts including pest control companies and wildlife protection agencies, the US Fish and Wildlife Service, the Humane Society, city animal control bureaus, and other professionals.

- **Lock up the temptations.** Your yard and garage are veritable cornucopias of food choices for these omnivorous animals. Cut off access to your discarded food by using trash cans with lids that lock, or lids that are too heavy for raccoons to open unaided (like bear-proof trash cans). If you can, put your trash away in a locked garage or shed, and only take it out on the morning of trash pick-up day.

- **Secure your compost.** Raccoons love meat, fish, and eggs, so put these animal products in your trash instead of your compost to keep these critters from trying to get inside. They may still take an interest if your compost attracts tasty worms and insects (which is inevitable), so keep your compost in a container that has a lid and a raccoon-proof latch.

If you don't compost but you keep piles of leaves and brush in your yard as a haven for birds, butterflies, and pollinators over the winter, the insects that enjoy and depend on these piles for survival may also attract raccoons. It's up to you if you want to clear out all of this very important bug habitat in favor of raccoon-proofing your yard; I can tell you that I maintain my piles of leaves and plant debris for the birds and insects, secure my trash, and take the occasional raccoon visit as it comes.

- **Keep pet food indoors**. If you feed your pets on your porch or in your yard, bring the food bowl in every night before dusk. Raccoons will help themselves to it nightly if it's within their reach, but pet food isn't good for them.
- **Harvest your fruit and vegetables promptly**. Fruit that falls from your trees is fair game to raccoons, so keep the ground under your trees cleared of fruit on a daily basis. Pick your fruit as soon as it becomes ripe to keep raccoons from beating you to it. The same goes for vegetables: While raccoons prefer sweeter fruits, they will sink their teeth into your tomatoes given the chance.
- **Keep raccoons outside**. It's easier than you think for these clever animals to find a way into your home. Any gap of more than a few inches can be an entryway for a young raccoon, so inspect your attic and basement and cover any holes with wire mesh, hardware cloth, or wooden boards. Look under your porch to see if raccoons have selected this sheltered area to settle down and raise a family. If not, secure every opening to discourage them from ever taking advantage of this secluded spot.

    Take another important step to keep raccoons from accessing your attic: keep your trees, shrubs, and climbing vines trimmed away from your home, so raccoons can't use these as a stairway to attic heaven.
- **Lock your pet door at night**. Your dog or cat may be free to let themselves out at night, but you don't want raccoons following them back into your living room. Latch this access after dark. Raccoon-proof pet doors may be the right solution for you; some

of these even read your pet's microchip to allow entry, keeping wild animals from passing through the opening.

- **Clean and store your grill**. Bring in utensils, grates, and any other items that came into contact with food while it cooked. Scrub these before putting them back outside if you store them there or keep them indoors until their next use. Like most animals, raccoons do not understand the concept of cooked food, but they know what they like.

If raccoons have already formed undesirable habits that involve your home and yard, try one of a number of ways to make your property less attractive to them.

- **Repel them with stink**. Raccoons hate the smell of ammonia or vinegar (can you blame them?), so soak a rag in one of these odiferous substances and place it wherever you don't want them to go. You will need to keep restoring this deterrent every few days or after a heavy rain. Commercial repellents are also available, though they can be costly depending on the size of the property you want to protect.
- **Get them wet**. Motion-activated sprinklers surprise and infuriate raccoons, driving them off for short periods of time.
- **Light them up**. An even better deterrent than sprinklers, motion-activated lighting will stop a raccoon in its tracks and make it think twice before entering your brightly illuminated yard. Raccoons are nocturnal by nature and don't like to forage where they can be seen. If the lights come on every time they visit, they will stop visiting.
- **Use night lights**. A bright light that stays on all night can discourage raccoons from ever approaching the lit area. Lights over your trash receptacle, compost bin, and porch could be enough to protect you from these midnight foragers.

Is It Rabid?

Raccoons can carry diseases including rabies, which will change their behavior and turn an otherwise placid animal into something very threatening. While a rabid raccoon is a rare occurrence, some communities see more of them than others. Know the signs and call your local animal control department as soon as you see something wrong.

- **Appearing in broad daylight.** Not every raccoon on the move during the day is rabid—in fact, most are just going from here to there—but this is the first clue that something may not be right with one. It's a cue to watch this animal for other symptoms.
- **Underweight and disheveled.** A sick raccoon probably has lost a lot of weight, and its fur may be rough and full of filth.
- **Aggression.** If a raccoon that doesn't know you personally comes running up to you, and if it attempts to bite you or hold onto you, get away from it as quickly as you can.
- **Stiffness.** Raccoons with rabies can have trouble walking and may lose the use of their back legs. This is an easy symptom to spot.
- **Foaming at the mouth.** Any animal that does this is in trouble. In raccoons (as well as foxes and dogs), this is one of the classic signs of rabies.
- **Lethargic behavior.** Most raccoons like to be on the move, so one that just wants to collapse is very likely ill.
- **Tearing at its own paws or skin.** This self-mutilation behavior is abnormal, and the resulting blood can carry pathogens.

Any raccoon may be carrying rabies without having any symptoms. Avoid physical contact with any raccoon, just as a good rule of thumb. If you do come into contact with a rabid animal, contact animal control immediately, and then contact your doctor. If you have been bitten or scratched, you will need to begin a course of medications and injections to counteract the illness. (If you wait until symptoms appear, it may be too late—and fatal.) Having animal control capture the suspicious raccoon can be important to your treatment as well; whatever hospital provides your care will want to test the animal to see if it is indeed rabid.

Alert, curious, and with a glint of intelligence in their eyes, foxes make a very good living in human neighborhoods. © PETER CLAYTON PHOTOGRAPHY, ISTOCK

Chapter 12

AS CLEVER AS THEY LOOK

FOXES

Wildlife in their backyard was one of the primary reasons that Luane and Peter Haggerty chose to live in a housing development that backs up to a small woodland. Deer frequent their yard and nibble on their trees and shrubs, opossums cross through on occasion, and rabbits visit to enjoy grassy expanses on a regular basis. So when an adult fox appeared one day and lingered in the backyard garden, the Haggertys considered it a special occasion.

"I thought it was beautiful, so red, gorgeous," said Luane. "But my neighbor freaked out. She put up a fence for her small dog."

No one in the neighborhood saw exactly where the fox and its mate chose to create their den, but after a couple of months, an adult emerged with three playful kits. "They were playing in my backyard," she said. "It was so regular, we put out dog toys to see if they would play with them—and they did. But they were more interested in each other."

Soon the next-door neighbor and her young daughter began watching them from her back porch. "You could set your watch by them," said Luane. "The puppies would appear every day right around 2:30 in the afternoon. We put out some treats—melon, birdseed, some dog treats, bones and things. They would find them and run all over the yard with them like maniacs, like they were saying, 'Look what I found!'"

Even while keeping a healthy distance from the playful kits, Luane and Peter had to adjust to the peculiarities of their new neighbors. Little foxes make sounds that can easily be mistaken for a crying baby. "Those are your foxes," Peter told Luane when she expressed concern for what might be out there. When the small dog next door left its enclosure and tried to cross over to the Haggerty yard to play with the kits, the dog's owner scooped it up and took it inside.

"They definitely ripped up the garden," Luane admitted, "but I definitely preferred them over the garden. It's all native; I don't plant things that are hard to take care of. But they ripped it up, or chewed it, or played with it—it was so much fun to watch them, I didn't care. I can see if you are a much more meticulous gardener, you might be sad. But it all grows back."

As the foxes grew, they took an interest in bigger prey, perhaps introduced to it by one of their parents. "One day it was all feathers in our backyard everywhere," Luane said. "It looked like a Canada goose. If I had to guess, I'd say one of the parents brought it to them. By the time we saw it, it was nothing but feathers."

When June came to an end, the kits had reached their adolescence and began looking for their own territories, so they no longer visited the Haggerty yard. The following year, however, the adults returned and produced two kits, and three more a year later. When other animals entered the yard, the Haggertys delighted in the interactions between species. "One of the foxes went up nose to nose to a groundhog and sniffed him, and the groundhog kind of punched him," she chuckled. "He just smacked him. So after that, the foxes left the groundhog alone."

The foxes did not reappear for a fourth summer, but the Haggertys have not ruled out the possibility that they may yet come back. "There are other places they may have gone right around here," Luane said. "We're on the edge of a development, and there's a Laotian temple right across the road. They've left everything wild there—it's like suddenly you're driving in Thailand. The foxes may have made a home in there somewhere."

For Amy Irwin, whose home is on the edge of a park with rolling hills in upstate New York, the sudden appearance of an adult fox on the pathway to their backyard shed stopped her and her husband Jerry in their car in the middle of the driveway. "I thought it was a statue. I almost said, 'Who put that fox statue in our yard?'" she said. "We just sat in the car and didn't want to get out. Then I said to Jerry, 'Look behind it.'"

From under the shed, several fox kits emerged and began to gambol about the yard. Nothing like this had ever happened before on their property, so they sat still and watched until they were certain that the kits had not even noticed them. "We went into the house quietly and went into the back room and watched from there," said Amy. "We could see them in the mornings. They would come out and play and roll around. I had a bird feeder full of seed, and they were out there eating it and rolling around. It was like having wild puppies."

The Irwins noted many of the same details that the Haggertys had enjoyed with their kits, particularly the sudden appearance of a "splot" of feathers, as Amy described it, signaling that a rock pigeon had been sacrificed for the kits' dinner. Eventually Amy tossed out a tennis ball for the little foxes to play with, and the kits rolled and pushed it around the yard for the Irwins' viewing pleasure.

"Then in a week they were gone," she said.

Both to encourage the foxes to move back into the den they had dug under the shed and to protect her pets, Amy began keeping her two free-ranging cats indoors, letting them watch "cat TV," the constant parade of birds, squirrels, chipmunks, and other critters through windows that overlooked the backyard. She did spot another fox one day, an adult that barked at her cat until Amy went outside and scared it off.

"Having the foxes was the highlight of our year," she said. "I hope they come back again sometime."

With its small size, bright chestnut coat, the bright white spot on the end of even the youngest kit's tail, and the undeniably intelligent look in its sharp face and bright eyes, the red fox appeals to humans in a way that many other four-footed wild animals do not. Even when it digs itself a den in an inconvenient spot, the fox's offspring can be so entertaining that homeowners willingly forgive them for the minor interruption in an otherwise orderly yard. Foxes see advantages to living in such close quarters with humans, from the ready availability of food to the protection human neighborhoods offer from larger predators. It's nice to know that in some small way, we can make these little red mischief-makers feel safe.

As Sly as Legends Say

Whether they live in the depths of a pristine mountain forest or under a redwood deck in a suburban subdivision, red foxes (*Vulpes vulpes*) find their way to a comfortable living. In fact, foxes have spread to six continents, making them the most widespread mammal in the world (after human beings), thanks to their ability to make the most of whatever environment they inhabit.

Part of this adaptability comes from foxes' varied diet. Not only do they eat mice, rabbits, and birds, but they also enjoy snakes, insects, and occasionally carrion, so they can make the most of whatever the nearest forest or field offers them. They supplement their diet with nuts, fruit, and berries, allowing them to plunder a squirrel's winter cache when much of its usual prey goes underground for the winter, or to feast on fruit ripening on vines or falling from trees in late summer and fall.

Foxes bring some distinct advantages to their hunt for prey: first, their hearing picks up low-frequency sounds like animals burrowing underground, so they can pinpoint the exact location of a tiny rodent and use its forepaws to dig with great enthusiasm at that spot. This remarkable talent receives considerable attention in nature documentaries about animal survival in winter conditions. Even under several feet of snow, foxes can still hear rodents scrabbling away or gnawing through roots and plant matter beneath the ground's surface. The fox stands very still to determine exactly where the tiny sounds of a mole, vole, or mouse may be; once it locates the sound, it leaps almost straight up, arches its back, and drops straight down headfirst through the snow to capture the little creature with one blow. In seconds the rodent becomes supper, making its way right into the fox's waiting jaws.

Not all of their prey stays underground all day, of course, so foxes apply another skill to chasing rabbits and other surface-feeding animals. Creeping as silently as possible to within close proximity of a feeding rabbit or mouse, the fox waits for the animal to take off at a run and pursues it at a blistering rate of thirty or more miles per hour. As fast as these rodents are, they may not be able to outrun a determined fox. Once a fox catches and kills its prey, it may not eat its conquest immediately, depending on how long ago it last ate. Foxes will hunt with a full belly and hide their prey for later, storing it in a snowbank in winter, under a pile of leaves, or in a burrow in the warmer months.

This brilliant adaptation to winter conditions probably began very early in the red fox's North American evolutionary history, when all of the continent's red foxes lived in the Northwestern region's boreal forests and glacial areas. In fact, these are the only native red foxes in the United States—the rest, according to legend and some historical fact, arrived once Europeans began to settle here and bring their love of sport hunting with them.

The story goes that Europeans who enjoyed hunting on horseback found colonial North America's indigenous gray foxes (*Urocyon cinereoagenteus*) quite unsatisfactory for pursuit. Gray foxes can climb trees, so the packs of hounds accustomed to following an animal's scent on the ground were utterly flummoxed by prey that could leave the ground altogether. To satisfy their passion for hunting in the time-honored way, these new American dwellers came up with a solution: bring red foxes over from the European forests to serve as more conventional targets. The transplanted red foxes established themselves quickly in terrain similar to their home country, so it didn't take long before far more of them lived here than these traditional hunters could pursue.

These foxes of the European subspecies are not the red foxes that went on to populate most of eastern North America, however. A recent study published in the *Journal of Mammalogy* found no genetic material from Eurasian fox species in the red foxes now living in the Southeastern United States. Instead, these scientists discovered that the Southeastern foxes are closely related to the foxes native to Eastern Canada and the Northeastern United States, making them the descendants of foxes that have been on this continent for hundreds of thousands of years. What happened, then, to the offspring of the European foxes? Common wisdom suggests that they were absorbed into fur farms, raised in captivity, and sacrificed to the clothing trade.

Today scientists can parse out nine different subspecies of red fox in North America—and forty-five subspecies worldwide, a testament to the *Vulpes* genus's ability to adapt to whatever its surroundings demand.

Even with all of these skills, however, foxes faced some monumental challenges in the early to mid-nineteenth century from their number one predator: humans. The figure of speech "like putting a fox in the henhouse" springs from true scenarios on family farms throughout North America, a synonym for placing someone obviously untrustworthy in a position of authority. Foxes eat chickens, so if one showed up near a family's chicken coop, only one solution would work: kill the fox. Hunters who supported their families by bringing in game animals—pheasant, grouse, turkey, and rabbit, for example—also objected to the numbers of foxes with which they had to compete for prey. In response, states established bounties for quantities of dead foxes, turning the practice of hunting them into a regional pastime— and for some, a good living.

As with all of the bounty schemes we've seen in this book, however, the resulting fox kill far exceeded what was necessary to protect farmyard chickens, or even to maximize the number of game birds and other animals hunters wanted for their own livelihoods.

"Most of the time the number of game birds and animals taken by foxes and other predators is insignificant compared to other natural losses," the Pennsylvania Game Commission tells us on the organization's website. "It's true that foxes take grouse, pheasants, rabbits, and other game, but these are usually 'surplus' individuals, those animals that would likely die from other causes—accidents, disease, starvation, etc.— before the next breeding season."

Trappers collecting foxes diminished the population even further. Most foxes that wind up in traps are first-year animals, just recently separated from the litter, so trapping these young foxes gave them no opportunity to produce even a first litter. The result: eventually, spotting the red fur and the bright white tail tip moving through a forest or field became a rare occurrence.

Lucky for foxes and a wide range of other fur-bearing animals, the lust for fur came to an end in the mid-1800s. Expensive and hard to care

As Clever as They Look: Foxes

Foxes often dig themselves a fresh den under a porch or garage, but they also may take one over from another animal like a groundhog or raccoon.
@ PORTSMOUTHNHCHARLEY, ISTOCK

for, furs became much less attractive to the clothing and haberdashery industries than less costly materials like wool and flannel. In the 1840s, the United States and the Dakota and Ojibwe tribes of the Midwestern territories signed treaties that ceded the tribes' hunting land to the United States, paying off massive debts the Indigenous people had built up with the federal government. With the stroke of a pen, the trapping era came to an end. Other regions soon followed suit, so by the 1870s, the European fur trade had nearly disappeared in most of North America. The remainder, driven by demand in Russia, shifted to northern territories in Canada.

Fur trapping continues today, with more than 176,000 licensed trappers still at work in the United States as of 2015. Their products are still in demand in parts of Europe, Russia, and China, but American fur traders face significant opposition to their industry from animal rights groups, so they maintain a generally low profile outside of their immediate business circles.

How to Live with Foxes

Seeing foxes' poise, finesse, and even their dapper appearance helps us understand why so many children's stories portray the red fox as a cunning animal, calculating its next move and carrying it out flawlessly. Take J. Worthington "Honest John" Foulfellow in Walt Disney's animated classic *Pinocchio*, a sly trickster who captivates the wooden boy and preys upon his dreams of becoming real; or the fox in Aesop's fable who cannot reach a juicy bunch of grapes, so he rationalizes away the grapes' palatability, declaring them sour and unworthy of his attention—an assessment he can make without even tasting them. Pagan mythology, fairy tales, and morality parables are stuffed with portrayals of clever foxes making the world around them spin in the direction they choose.

It's no wonder, then, that we instantly expect foxes to . . . well, *outfox* us, even in our own backyards. This fairly small creature—a large red fox weighs in at about eleven pounds—has convinced us that it nonetheless has a big range of tricks at its disposal, from its ability to dig its way under just about any structure to its amazing capacity for seeking out and catching its prey, even when its target is underground.

Foxes hunt in daylight, so seeing one in the middle of the day is not a cause for alarm. Despite being fairly well integrated into human neighborhoods, they continue to be very wary of people, so you have little to fear from a fox crossing a backyard or dwelling in the "forever wild" area behind a home. If a fox takes up residence in your yard, or if you have small pets that may be attractive prey for a fox, you can take some steps to keep them away from your home and to discourage them from returning.

- **Keep accidental food sources out of their reach**. Lock up your trash, either by storing your receptacles in a garage or shed or by securing the lids. Feed your pets indoors or under supervision if you need to feed them outside. Don't leave bowls of food outdoors unattended.
- **Keep your small pets indoors**. If you have a kitten, rabbit, guinea pig, bird, hamster, gerbil, or any other small animal, do not leave it outside unattended. A domestic rabbit in a backyard is a wide-open invitation to a fox (or a number of other wild animals) to make a meal of your fur family member, even if you are close by. Cats and dogs that nearly equal a fox in size are much less likely to become fox targets.
- **If possible, keep your animals indoors**. A shed or barn may be the better option for keeping small livestock safe from foxes. While foxes generally don't confront cats or dogs that equal them in size, your pets may approach a fox, which can lead to a row and injuries. Keeping these pets inside when you can't keep an eye on them yourself will virtually guarantee that no such battles take place.
- **Scare it off**. It doesn't take much to keep a fox from hanging around your yard. Leave a radio on in your backyard or install motion sensors that set off an alarm, a sprinkler, or bright lights. If you see a fox while you are outside in your yard, toss a tennis ball in its general direction. These little annoyances are enough to startle a fox and convince it to look elsewhere for its next meal.
- **Try a dog repellent**. Garden centers and hardware stores sell products that keep other people's dogs out of your yard. These work well with foxes, too, and are readily available when you need them.

If a fox has chosen an inconvenient area of your property to dig its den but does not yet have kits, you have a few options for encouraging it to choose somewhere else. These methods are considered "mild harassment," and will simply make the fox uncomfortable.

- **Foxes hate smelly stuff.** The same strong smells that you don't like also bother foxes: sweaty gym clothes, cat urine in kitty litter, or a vinegar or ammonia-soaked rag. Place these near the opening of the site the fox seems to have chosen, and the stink may be enough to drive it off.
- **Try mylar.** Get some reflective balloons and tie them so that they bounce in the breeze not far from the den opening. If balloons are too big for the space, strips of mylar ribbon blowing in the wind can be scary enough to prompt the fox to relocate.
- **Spicy repellents work for foxes.** Foxes don't care for capsaicin, the extract from chili peppers that brings the heat. The smell alone may repel them enough to make them abandon a chosen den. Hardware stores carry a granular compound spiked with plenty of capsaicin (also known as capsicum); spread this near the den opening.

On the other hand, if kits have been born in the den, your best course of action is to simply wait for nature to take its course. When the kits have grown enough to leave the den's comfortable shelter, they and their parent will move off on their own. Once they are gone, double-check to be sure that there are no kits remaining before you take any further action to keep foxes (or any other animal) from using this spot again.

- **Bury a barrier.** Foxes dig to get where they want to go, so you'll need to bury hardware cloth or another impenetrable material about a foot down to exclude them from entering again. If you have other places around your property that might be attractive to foxes in the future—a shed, deck, or porch that they can access by digging their way underneath—install the same kind of barrier around the perimeter of these structures.
- **Keep them out of other animals' enclosures.** If you have a chicken house, dog kennel, or other place where animals you actually want in your yard are left unsupervised, you can protect them with the same kind of barrier around the outer perimeter. Adding

a low electric fence—a single strand of electrified wire about four inches above the ground, about twelve inches in front of an existing fence—will deter foxes and other animals from approaching your outdoor livestock or pets.

American alligators do not actively hunt humans, but they see anything in the water as fair game for an easy meal. © NIC MINETOR

Chapter 13

ASSUME THERE ARE ALLIGATORS

Kevin Albert Murray, a forty-one-year-old resident of Northport, Florida, had just finished mowing a lawn in Port Charlotte on July 15, 2005, at about 7:00 p.m., when he decided to take a quick swim to cool off. The Apollo Waterway's canal near his client's yard looked inviting, its clear water showing just a bit of current as it shimmered in the sunlight. As he had done on many other hot July days, he stripped off his shirt and jumped off the dock, immersing himself in the water up to his neck.

Perhaps Murray had heard from neighbors there that an alligator lived along this stretch of canal, but none of them seemed to think that the twelve-foot, 1,200-pound alligator presented a threat to life and safety. The animal just laid there most of the time, sunning itself on the grass along the canal's edge. Alligators are opportunistic feeders, rarely feeling motivated enough to chase prey—nor do they have to, as most of the sustenance they need swims into their mouths as they wait, concealing their presence by blending seamlessly with mud, rocks and silty ponds and canals.

He didn't see the massive reptile until it sank its teeth into his right arm and thumb. Murray screamed, as any sane person with alligator teeth in his forearm would do, and thrashed as hard as he could as the massive animal pulled him underwater. Someone who heard him cry out or saw him go under called the Charlotte County Sheriff's Office at 7:17 p.m., and at 7:40 p.m., state trappers Tracy Hansen and Bo Davis reached the spot where Murray had gone into the water.

They didn't have to search long. The giant alligator came to the surface on the opposite side of the canal, its jaws still clamped around Murray's right arm as it dragged his lifeless body into some bushes.

"The alligator's first instinct when it's attacking is to spin and drown the victim," Davis later explained to reporter Kristen Kridel in the *Ocala Star Banner*. The chances of finding Murray alive, then, had been slim to none from the moment the animal attacked him.

Reinforcements arrived as Hansen and Davis began to move, and the two additional officers became part of what was now clearly a recovery plan rather than a rescue. The four got into a boat and crossed the canal, edging as close as they could to the alligator without driving it off. When the twelve-foot reptile saw them coming, it released Murray's arm and let his body float away. The officers retrieved Murray from the water, confirmed that he had drowned when the alligator held him underwater for at least twenty minutes, and brought his remains to shore.

Recovering Murray's body was the easy part. Now they had to capture the alligator, sedate it, and get close enough to destroy it.

They began this monumental task by attempting to bait the alligator using two cow lungs on ropes held just over the water. The alligator showed no interest and vanished under the surface. This actually gave the trappers an advantage: they got out flashlights and turned them on, shining them into the water—and in a few seconds they saw the alligator's eyes glittering back at them. Now they knew they were close enough to try a more effective tactic, one involving a skill unique to animal control officers in the Southeastern United States.

Hansen brought out a crossbow and took aim just past the animal's glowing eyes, then let the tethered dart fly. The tranquilizer dart hit its target and pierced its leathery hide, lodging there securely enough to allow the officers to use the attached rope to reel in the alligator. They pulled it to within four feet of the boat, where one of them put a gun to the animal's head and pulled the trigger.

But "the bullet only stunned him," Hansen later told Sarah Lundy of the *Fort Myers News-Press*. Nonetheless, it gave them the opportunity they needed to tie a noose around the disoriented alligator's head and start pulling it toward the shore. By the time they reached the bank, however, they had a new problem: the tranquilizer had started to wear off, and the alligator began to struggle against the noose.

They quickly put a stop to this with another bullet.

This time the shooter killed the reptile. They loaded the beast into a truck and took it to Lakeland for a necropsy (the animal equivalent of an autopsy), and to examine its stomach contents to determine if it had actually consumed any part of Murray. The alligator capture/kill operation had lasted just about an hour.

Remarkably, officers told media outlets that day that they had never received a call about a nuisance alligator along the Apollo Waterway, so this attack took them by surprise. Residents of the Southeastern states

Hidden in the brush alongside a pond, an alligator waits for unsuspecting, incautious prey to approach and swim close enough to grab with its jaws. © NIC MINETOR

where alligators thrive—Florida, Alabama, Mississippi, Louisiana, the Gulf Coast of Texas, Georgia, and South Carolina—tend to be a little blasé about their big reptile neighbors, seeing them sunning next to a pond or gliding across a canal while indicating no interest at all in the humans around them. Alligators are so common across the coastal wetlands of these states that there would be little point in panicking over them, but they can't be ignored either: in Florida alone, from 1948 to 2021, the Florida Fish and Wildlife Conservation Commission (FWC) recorded twenty-six fatal attacks by alligators out of 442 unprovoked bites, 303 of them severe enough for the victim to be hospitalized. That's an average of six bites per year and a fatality every three years—certainly not enough to paralyze a community, but enough to prompt residents to use caution when approaching water features.

What You Need to Know About Alligators

The state of Florida hosts about 1.3 million alligators, none of which are staring at you with eyes just above the water and thinking, "I'm going to leap out and grab that." These large reptiles eat animals that are easy prey and within their immediate vicinity—exclusively animals that are smaller than they are: fish, turtles, birds, and small mammals, as well as some smaller snakes, creatures that they can overpower without much of a fight.

When they hunt, they usually do so in water. So, your best defense against them is simple: stay out of bodies of water that may contain alligators.

The FWC keeps records of every known human encounter with these animals, and their statistics are surprising: Even with the significant human population growth in Florida in recent years, there has been no upward trend in alligator attacks. "The likelihood of a Florida resident being seriously injured during an unprovoked alligator incident in Florida is roughly one in 3.1 million," the FWC assures us. For the most part, unprovoked incidents have occurred in residential areas with ponds or canals, or in parks where people swim or snorkel.

Assume There Are Alligators

Many of the Southern states have seen growth in residential development in the past few years, as retirees from the Northeast begin to migrate to warmer climates. As open land turns into housing tracts developed around water features like ponds and canals, more alligators and humans will come into conflict. When new housing and commercial developments are built, contractors create basins and trenches from which soil comes for construction. Turning these areas into stormwater retention ponds or canals is a legal requirement for building new housing tracts, so these bodies of water then become prime habitat for alligators. And if the ponds and canals connect with existing creeks, lakes, and other waterways, the alligators have an easy transportation network to explore; they move into areas that are not already dominated by large alligators, establishing their own territory.

These reptiles present one of the best illustrations of people moving into an animal's prime habitat and taking it over, leaving that animal with only one option: cohabiting with humans. Virtually every suburban neighborhood in Florida has its own ring of streets and lawns that end at the edge of a canal, creek, or small lake. Here alligators lounge as ominous curiosities, paying little attention to humans until one of them takes a swim—and then it's simple for an alligator to mistake a human hand or foot for a fish and move with surprising speed to grab it. In many cases, the unfortunate victims were swimming in seemingly isolated ponds or lakes, unwittingly presenting themselves as easy targets.

The most notorious case of this took place at Disney's Grand Floridian Resort in June 2015, when an alligator attacked two-year-old Lane Graves as he played in the sand at the edge of a pond. Several other guests had seen an alligator swimming in the pond, and one photographed it from the balcony of his hotel room and then started down to the beach to warn people there. Just as he set foot out of his door, he heard Lane's mother scream.

In an instant, the alligator had come out of the water and grabbed the toddler's head between its jaws. It dragged Lane underwater and held him there—lifeguard Christopher Tubbs, a Disney employee, told the investigating officers that he caught a glimpse of Lane's feet just barely out of the water about twenty feet from shore as the reptile swam away with him. The child perished in the pond.

"The alligator may have had a diminished fear of people by being in an area with lots of humans," the Associated Press (AP) reported when the FWC released its final report on the incident a year later.

Disney responded to the attack by building a stone wall around the lagoon, adding signs that say, "Alligators and snakes in area" and "Stay away from the water," and eliminating fishing from shore, limiting it only to excursions, the AP and CNN reported.

Alligator Challenges and Recovery

Alligator skin makes a particularly desirable and valuable leather, so hunters in the 1800s worked tirelessly to harvest these reptiles and bring their skins to manufacturers. Like so many of the trapping and killing operations we've seen for taking the hides of animals throughout this book, however, this zeal ended up depleting the alligator population so severely throughout the Southern states that the federal government had to take steps to curtail it. When annual harvest figures in the early 1950s dropped precipitously, it became clear that alligators had been hunted nearly to extinction.

Alligators faced other challenges as well. In the 1950s, pesticides like DDT (dichloro-diphenyl-trichloroethane) reduced alligators' ability to produce viable eggs, curtailing their reproduction just as overhunting began to deplete their numbers. Pollution in waterways from oil spills and corporate dumping of chemicals threatened the rivers and lakes where they lived, and the explosion in Southern states' popularity—both as a resort destination and as a place to live with year-round warm weather—meant that sprawling construction subsumed the region's wild lands.

Alligators needed human rules and regulations to come back from the brink of obliteration, and luckily, they received exactly that. In Louisiana, the Department of Wildlife and Fish (LDWF) shut down alligator season from 1962 to 1972, allowing the reptiles to reproduce at normal rates and stabilizing the population. The state developed a wild alligator management program that included saving the animals' wetland habitat, conserving the alligators while providing financial incentives to landowners to protect wetlands on their property. Nowhere has this been more successful than in the gorgeous Cameron Prairie National Wildlife Refuge, turning former commercial rice fields into protected habitat for alligators, wading birds, waterfowl, and other water-loving animals, while still permitting hunting on a limited basis. Preserves like this one sprang up all over the state, increasing tourism from birders and wildlife lovers while allowing these animals to thrive.

When the Endangered Species Act passed in 1973, the American alligator was one of the first species listed for federal protection. Louisiana led the region in creating an alligator management plan that includes an annual controlled wild harvest and an alligator ranching program, treating the animals as a renewable resource by releasing 10 percent of the farm-raised alligators into the wild, while selling the rest for food or use of their hides.

Programs like this one helped to increase alligator populations across the Southern states, so much so that the US Fish and Wildlife Service removed the American alligator from the Endangered Species list in 1987. The international trade in alligator hides continues to be regulated to prevent another drop in the species population. Today, their numbers are now at a healthy five million across all of their resident states, up from just one hundred thousand across ten states in the 1950s.

DEATH BY ALLIGATOR: WHAT YOU NEED TO KNOW

Another of the deaths caused by an alligator attack illustrates a very important point about these animals: the messy ways they can kill a person. An alligator attack that took the life of fifty-four-year-old landscaper Janie Melsek led to an overhaul of alligator policy on Florida's Sanibel Island.

As Melsek quietly trimmed vegetation around a backyard pond on Sanibel, an eleven-foot-nine-inch alligator rose up out of the pond, grabbed her right arm, and dragged her into the water. She fought the 457-pound animal and would have been killed on the spot had it not been for the quick action of a neighbor, Jim Anholt, who heard her screams and ran over to help. He called police and held her head out of the water, doing his best to resist the alligator until they arrived.

"It was kind of a tug of war," Anholt told the *Fort Myers News-Press* later. Three police officers joined him and struggled with the alligator for five minutes, finally winning the battle and pulling Melsek out of its jaws.

By this time emergency medical technicians were on the scene as well. They began controlling her bleeding and treating her wounds. The bleeding was severe, but another threat made this a critical case: the risk of infection from alligator bites. Many victims of alligator bites develop systemic inflammatory response syndrome (SIRS), which can lead to organ failure, shock, and death in very little time.

Melsek's injuries were so severe that doctors amputated her right arm below the elbow. She also suffered bites on her buttocks, back, and inner thighs, making it unlikely that she would escape the body's inevitable response to severe alligator bites. Two days after the attack, even as trauma surgeons worked to combat the spreading inflammation, she died.

Police, meanwhile, took the opportunity to shoot the alligator in the head. Even in death, the alligator put up a struggle, its sheer size making it tough to remove from the pond. Six men worked to pull the dead animal to shore and put it in a truck, sending it on to the FWC for a necropsy.

This tragedy touched off a tinder pile of anger and fear among Sanibel residents. The island had its own set of rules for dealing with alligators, protecting any alligator over four feet long unless they became aggressive—usually because tourists fed them, causing them to lose their wariness of humans and to associate people with food. After such a large alligator killed Melsek, however, the city council agreed that their policy had to be broadened to provide much better protection for their human residents. Two weeks after the landscaper's death, the council voted unanimously to allow the city to hire trappers to kill an alligator more than four feet long "if it makes people feel unsafe." This brought the island in line with rules established by the FWC, which the rest of the state already followed. In the next two months, residents reported forty-four alligators that needed to be removed from their neighborhood, and trappers and police made quick work of them.

How *Not* to Get Attacked by an Alligator

I covered quite a number of other such attacks in South Florida in my book *Death in the Everglades*, if you have an appetite for more of these gruesome accidents. For the purposes of this book, and especially if you live in one of the ten states that have resident alligator populations, it's more important that you come away with a healthy respect for the alligator that may be lounging in your backyard or along a nearby canal at this very moment. Here is what you need to know to avoid becoming a reptile casualty in the future, prepared originally for my Everglades book.

- **Leave alligators alone**. Generally, an alligator will not approach you. No matter what you may have seen in TikTok videos or selfies, these shy creatures see no benefit in a relationship with human beings.
- **If the alligator's hissing, you're too close**. Maintain a safe distance of at least fifteen feet from the animal. If you are at least fifteen feet away and it's still hissing, back off even farther.
- **Do not harass alligators**. It seems ridiculous that this needs to be said, but do not attempt to tease an alligator, kick it, touch it, throw things at it, or poke it with a stick. Harassing an alligator is a federal crime, so on top of risking life and limb, you can be arrested for behaving foolishly with one. While they may look slow and lazy, alligators can move very fast when provoked.
- **Pay attention to your surroundings**. Keep an eye out near freshwater features like canals, ponds, streams, and lakes. Avoid areas of

dense vegetation around water. If you see an alligator, put distance between the alligator and you as quickly as possible.

- **Do not feed alligators.** It is against federal law to feed a wild alligator. Alligators that take food from humans come to associate humans with food and become more aggressive around people. Do not contribute to this.
- **Skip the selfie.** Turning your back on an alligator and inching in closer for a selfie is a bad idea. While there are no actual cases (that I know of) of a person being attacked from behind by an alligator while taking a selfie, there is no way for the animal to know what your intentions are when you make this potentially threatening move. National and state parks have plenty of stories of people being gored by buffalo, elk, bighorn sheep, or moose when trying to get in close for a photo. Don't take the chance on becoming a viral video for all the wrong reasons.
- **If you fish, throw your scraps in a trash can.** When you leave fish scraps around, you attract alligators, which may then become accustomed to hanging around where fish scraps are plentiful. This amounts to feeding an alligator, which can have dire consequences.
- **Swim only where it's safe, like in a pool.** If an area has signs posted that warn of possible alligators, it means that one or more alligators have actually been sighted there. Do not swim or enter the water where you see these signs. An alligator can mistake your hand or foot for a fish and try to catch it, resulting in a severe bite at best, and, at worst, the loss of a limb and a potentially deadly infection. It also may hang on and drag you underwater for the express purpose of drowning you.
- **Swim in daylight.** Alligators are more active at night, so stay out of the water after dark. Again, stick to a manmade pool for your evening swim.
- **Never leave children unattended.** Do not let your children run around near water or play on the edge of a pond or stream with-

out supervision. A tiny hand or foot in the water can be seen as prey by an alligator. A lack of a warning sign at a pond does not mean that this pond has no alligators.

- **Keep your pets away from alligators**. Alligators may attack small dogs and cats that they know they can overpower quickly. Walk your pet some distance away from a pond or stream, even if there is a paved path along the water.
- **If an alligator charges you, run away in a straight line**. Do not run in a zigzag pattern, as this makes it easier for the alligator to catch you. Run directly away from the animal.

We have less to fear from the striped skunk's legendary defense mechanism than most people realize, as they only deploy it when they are truly threatened. © NIC MINETOR

Chapter 14
FRAGRANT NEIGHBORS
SKUNKS

My mother harbored an intense fear of all animals, from the German shepherd police dog living across the street to the tiny field mouse she once spotted dashing out of the wildflower meadow behind our house and into our yard. We never owned a pet, except for one very sad budgerigar parakeet that she never allowed to leave its cage, leading it to develop a depression so severe that we had to rehome the poor bird with my animal-loving cousin. Dogs, cats, rabbits, moles, gerbils, hamsters, bees, crickets, ants, even the tiny pet turtles from Woolworth's gave her such willies that she had to run out of the room if any critter entered it. I have no idea what caused this near-psychosis level of terror, but it hampered many aspects of our lives—especially for me, an outdoors-loving kid who only wanted to go outside and play with whatever wandered into our yard.

So when the skunk showed up inside the chain-link fence around our backyard one sunny May morning in 1965, I was not certain if we should call animal control or the paramedics. Mom went into an apoplectic state that made further action nearly impossible.

The skunk, alarmed by its sudden confinement and clearly unable to find whatever tiny hole it had shimmied through to get into the yard, followed the inside perimeter of the fence counterclockwise over and over again as it searched for an exit. This brought it into closer proximity with our kitchen's sliding glass door, where Mom sat at the table in a nearly catatonic state, the phone receiver in her hand, moaning and breathing hard. "What do we do? What do we do?" she asked over and over, gasping for breath as the skunk passed by the concrete step just under the sliding glass doors directly in front of her. Finally, she dialed 0 to reach an operator, the only number she seemed able to summon to her consciousness.

When the operator answered, Mom blurted, "We have a skunk in the backyard! You have to help me!"

As if she heard this kind of panic every day, the operator responded, "I will connect you with the police."

A moment passed, and Mom "gave a *geshrei*," as we said in Yiddish in our house, meaning that she wailed an "Oyiiiiii" while waiting for someone to answer. Finally, after what seemed like an hour but was probably less than a minute, a man's voice answered. "Can I help you?"

"There's a skunk in my yard. A skunk!" Mom answered.

"Oh, wow, that sounds like fun!" the policeman said with a chuckle.

"Fun? Fun? What do you mean? I have to get it out! What if it does what it does?"

"You mean spray? It probably won't."

"You don't understand. We have a fenced-in yard. I can't go out there. I just can't!"

Now the police officer seemed to get what was up with Mom. His tone changed. "Ma'am, what is the skunk doing now?"

"It's following the fence around, circling the yard, over and over."

"Okay, ma'am, does your fence have a gate?"

"Yes," she breathed.

"All you have to do is open the gate, and the skunk will find its way out."

"Are you sure?" she asked. "What if it doesn't go?"

"It's looking for a way out now. It will find the opening and go."

"But what if we go out there and it sprays us?"

"Just wait until it isn't close to you and then open the gate."

This must have been the logic Mom was waiting for, because she thanked the officer and replaced the receiver in its cradle. She turned to me, her voice very grave. "All right, Randi, you have to go out there and open the gate."

"Me?" This felt like a very adult assignment all of a sudden, and I was seven.

"What, do you think *I'm* going to do it? It has to be you."

I looked at her widened eyes and her hand clutching the upper placket of her blouse as if the skunk were going to burst through the glass door, clamber up onto the table and leap into her shirt, and I knew that she was right. I had to step up.

I went out through the garage, walked around the edge of the building to the gate, and lifted the latch. The clank of metal against metal startled the skunk, now on the other side of the yard, so that it stopped still and turned to look at me. With one deep breath for oxygenated courage, I pulled the latch back and kicked the gate open so that it swung wide away from me, and I bolted for the garage.

Standing in shelter and watching out the garage window, I caught a glimpse of the black-and-white striped animal coming through the gate, rounding the corner of the fence and dashing off into the meadow behind it.

We were saved, and I was a hero. But were we ever in any real peril?

The Humane Society of the United States says no—and even if we were, the skunk would have given us fair warning. "Because skunks are generally easygoing, they will not intentionally bother people," the website tells us. "In fact, skunks may benefit humans by eating many insects and rodents many regard as pests."

Skunks do have a surefire defense mechanism that repels predators and non-predators alike: when they are frightened, they can release an oily spray from their perianal glands with a very unpleasant, quite distinctive odor, not unlike some strains of modern marijuana. "Those who have never smelled it may realize some of its power if they imagine a mixture of perfume musk, essence of garlic, burning sulphur [*sic*] and sewer gas, intensified a thousand times," noted Ernest Thompson Seton in his seminal 1909 book, *Life Histories of Northern Animals, Volume II: Flesh-Eaters*. "It is so strong that under certain circumstances it can be smelled for miles down wind."

This spray, however, supplies the added hazard of pain and burning if it gets in your eyes. Skunks only emit this oil when they believe that they and their young are being threatened by a predator, so most people who encounter a skunk have nothing to fear if they keep their distance.

The spray itself consists of thiols, volatile compounds that evaporate quickly and leave only the foul-smelling odor. This particular compound contains sulfur, much like human flatulence—but this is weapons-grade sulfur that lingers in the air like the rotting of spoiled vegetables. A skunk's perianal glands contain roughly two teaspoonfuls of this stuff, but they can project this with a force that spreads it up to twenty feet with extraordinary accuracy. If you don't approach the skunk and make sure you're more than twenty feet away from it, you won't get sprayed even if it does perceive you as a threat.

On the off chance that you've frightened the skunk, it will let you know whether it feels threatened by you before it fires with both barrels. The skunk will stamp its front feet, raise its tail, and hiss at you, and will pretend to charge you by jumping forward. This is your cue to get away from the animal before it actually turns around and points its rump and tail at you, the last sign before it sprays. If you've got a skunk dancing in front of you, get away from it and let it be.

Perhaps even more important, if you know you have skunks in your neighborhood, keep your dog leashed when you walk with it and don't let it approach a skunk. Dogs don't understand all of these warning signs and are likely to continue to try to approach, bark, or even engage the skunk in play—any of which may result in the most unpleasant shower your pet has ever had.

A Brief History of Skunks

Striped skunks (*Mephitis ephitis*) live just about everywhere in the United States today, with a remarkable thirteen different subspecies separated by geography. The average viewer may see little to no difference among these varieties, and indeed, the disparity is in the details: one may have a thinner tail, while another's skull is just a bit larger, and another has wider white stripes on its back. Some of the subspecies have appeared to make evolutionary modifications based on the area they call home: the Great Basin skunk (*M. m. major*), for example, has longer hind legs and feet, presumably for navigating tougher terrain than some others. Louisiana skunks (*M. m. mesomelas*) are smaller with shorter tails, perhaps to help them escape predators and slip into smaller holes in the featureless rice fields and bayous.

Generally, however, North America's striped skunks are easily identified by what they have in common: a black coat with distinctive white stripes down the head, back, and tail; short legs, keeping them close to the ground; a luxuriously furry tail, and a small, cone-shaped head. The far less prevalent spotted skunk (*Spilogale*) has more white stripes than the striped skunks and may have some white spots as well, depending on which of the four subspecies you encounter, but these are much rarer in urban areas despite their wide distribution across the continent.

Rarely more than thirty inches long and weighing about twelve pounds, skunks do not present a fearsome presence—in fact, were it not for their legendary defense mechanism, we would be hard-pressed to find these animals the least bit threatening.

Skunks have inhabited North America for at least 1.8 million years, the age of the first skunk fossil discovered in the famed Broadwater archaeological site in Nebraska. Originally residents of southern North America, skunks moved northward as the Wisconsin glaciers receded from 10,000 to 4500 BCE, eventually inhabiting just about every corner of the continent.

Indigenous peoples trapped skunks to make robes and coats of their fur, and this practice continued as European trappers arrived—but despite the desirability of their pelts, skunks did not face the same demand from the fashion industry that nearly wiped out many other mammals at the height of the fur trade. The main reason for this came from farmers, who found that skunks served as valuable partners in reducing the quantities of crop-eating insects that threatened their livelihood. A 1923 report generated by the US Department of Agriculture validates this argument, noting that despite claims that skunks plunder poultry farms and eat chicks and eggs, field experts with the nation's Biological Survey recorded the contents of sixty-two skunk stomachs and found that "grasshoppers and crickets formed a large percentage of the food of nearly half the skunks examined. Beetles and their larvae formed the most important items of food . . . and in many instances being their solo diet."

Skunk kits, born in a den under a shed in this front yard, may play and explore for weeks before moving off to find their own territory. © NIC MINETOr

When farmers received this confirmation of the skunks' complementary role in their endeavors, they demanded legislation to protect the animals from the fur trade. Skunks have continued to live on or near farmland since then, enjoying the protection of agriculture and devouring all the destructive insects they can hold, including such annoyances as army worms, tobacco worms, white grubs, hop grubs, grasshoppers, cutworms, cicadas, crickets, sphinx moths, and just about every other bug that farmers hate.

In addition to this devotion to an insect diet, skunks are actually omnivores, digging into burrows and eating rats and mice, as well as bird eggs, snake eggs, lots of plant matter, and the feces of other animals. This, along with the ever-present potential for stink, makes some homeowners in suburban areas see these striped coresidents as pests, bent on destroying their gardens and lawns. When the local skunks devour the carefully cultivated vegetation around their houses, it's hard to keep in mind that they also eat grubs and other larvae gardeners find deplorable.

And then their unsuspecting dogs, thinking they've found a backyard playmate, approach a striped skunk and have no ability to read the signs that the black-and-white visitor is getting agitated ... and the dog comes home reeking of skunk spray. By the third or fourth time they have to douse their pets in hydrogen peroxide, baking soda, and dish soap (see the recipe later in this chapter), homeowners have developed a very low opinion of their resident skunks.

Without question, skunks have moved into urban areas in numbers never seen before, and some cities have had enough of it. *Outside Magazine* writer Christopher Kemp wrote in 2014 that in his hometown of Grand Rapids, Michigan, a "thick, immovable cloud of skunk odor" blanketed the city as the summer heat set in, and he offers anecdotal accounts of neighbors who have reached the limits of their tolerance for this. MLive, Grand Rapids' Fox network affiliate, refuted this days later by pointing out that Grand Rapids does not have a municipal pest control program, and instead sends its residents with skunks on their property to private pest control companies, so the skunk "problem" seemed much greater and more costly to homeowners than it may be in actuality. Indeed, a skunk control industry seems to have blossomed there organically, with a simple Google search turning up more than a dozen providers. Similar trends have come to light in cities in Indiana, Ohio, and Illinois, making the Midwest an urban skunk nucleus of sorts.

THE RABIES PROBLEM

This surge in skunks sagged shortly after that, however, with an influx of rabies that reduced the overall population. Wildlife experts have come to expect a cyclical rise in animal diseases like rabies and distemper as a natural way to reduce the overall population of a species. A spike in such illnesses every five years or so keeps skunk numbers in balance with the available food sources, so while an increase in rabid skunks (or bats, foxes, or raccoons) can create greater hazards for people, livestock, and pets, this is not a cause for panic. In fact, such a spike may be in effect as I write this: in December 2023, Michiganders were alerted by their state Department of Health and Human Services that a local skunk breeder may have allowed rehabilitated wild skunks to comingle with captive animals, an error that led at least one skunk sold as a pet to contract rabies. "It can take months for rabies to show up in skunks," the state noted, reminding residents that once symptoms show up, the disease is always fatal.

The Indiana Department of Health reported in February 2024 that skunks with rabies had been detected in two counties in the southern part of the state near Kentucky, in an area where rabies is rarely if ever detected. "These are the first cases confirmed in Indiana since 2004," the news release said. Adding more evidence, the *Minnesota Star-Tribune* reported in August 2024 that rabies cases in skunks had risen dramatically over 2023, as evidenced by the numbers of diseased and dying cattle. Rabies spreads through contact with the saliva of a diseased mammal, usually one emboldened and made reckless by the infection; so a skunk with its judgment impaired by disease may approach and bite livestock, dogs, cats, or people.

The Centers for Disease Control and Prevention (CDC) operates the US National Rabies Surveillance System and produces disease reports on an annual basis. For the year 2022—the most recent year available as of this writing—more than 3,500 cases of animal rabies were reported, with Texas, Virginia, Pennsylvania, New York, North Carolina, and California leading the states with the highest numbers of reports. Skunks represented 18.4 percent of the total number of cases, or 660 reported rabid animals. "Skunks are a rabies reservoir across most states in the Midwest and Western parts of the US," the CDC tells us. "While exposures to rabid skunks are not very common in the US, when they occur >20% of skunks that expose people or pets have rabies. This means that skunks are the highest risk for rabies in the US, if they bite or scratch a person or pet."

The good news is that US mortality rates among humans who contract rabies are the lowest that they have ever been. Back in the 1960s, hundreds of people died of rabies every year, experiencing a decline over just a few days as the disease progressed, moving from flu-like symptoms to neurologic disorders: anxiety, confusion, hallucinations, disorientation, episodes of terror and/or excitement, fear of water, seizures, even paralysis, before finally lapsing into a coma shortly before death. Today, thanks to public awareness of the need to act quickly and vaccines that can prevent the disease from developing, less than ten people die of rabies annually.

Still, no one wants to be one of those ten people. If a skunk in your yard seems disoriented—walking in circles or stumbling, for example—or if it approaches you and charges at you as if to attack, it may be rabid. Foaming at the mouth is almost always an indication of rabies as well. Do not touch or approach the skunk; bring your family and pets indoors and call your local professional animal control organization immediately.

How to Live with Skunks

Seeing a skunk cross your yard is no cause for concern, but it may be a clue that you need to make some changes to keep them from selecting your property for their den (if that's something you would find inconvenient or potentially hazardous to your pets). You will find that the rules that apply to raccoons, foxes, and other wildlife also apply to skunks, as they all enjoy exploring your trash or compost for an easy meal.

- **Secure your trash.** Like so many other urban-dwelling animals, skunks are opportunistic feeders that will be delighted to pick through your accessible garbage. Keep your trash receptacles tightly closed or lock them up in your garage or shed when they are not on the curb awaiting pickup.
- **Feed your pets indoors.** If you must feed your dogs and cats on the back porch or in an outdoor enclosure, bring in this food at night to keep skunks from plundering it. Once a skunk knows there's food on the porch, it will return night after night to enjoy it.
- **Cover your window wells.** An easy source of water for animals, wells around your basement windows can attract lots of different kinds of wildlife. Place screen over these so they still have ventilation and fasten this down to keep inquisitive animals from breaking into them.
- **Don't overwater your lawn.** Skunks are grub eaters, so a saturated lawn will push these tasty morsels close to the surface. This is an

easy meal for skunks, especially if you water your lawn on a timer after dark, even when it may not actually need watering.

- **Keep your outbuildings closed and locked.** Skunks may find your open garage or shed irresistible, particularly if either of these contains trash that looks yummy to them.

How can you tell if a skunk has selected your yard for its den? Here are some signs and tips for determining if a skunk has moved in:

- **Check the old dens.** If a fox or groundhog has raised a family somewhere on your property, a scavenging skunk may take the path of least resistance and move right into that abandoned den.
- **Watch where it goes.** If you're seeing a skunk in your yard regularly, keep an eye on it to see if it's making its way under your deck or porch, through a hole at the base of your shed, or back to the same brush pile again and again. It also may burrow under an outbuilding or even under a cement slab patio.
- **Watch for tracks and droppings.** It's easy to follow tracks in mud or across your lawn to see if they lead you right to the den. Keep an eye out for animal droppings as well—if you don't know what skunk droppings look like, I highly recommend the book *Scats and Tracks of North America* by James Halfpenny (Falcon, 2019), which will help you sort out which animals are trekking across your property.
- **Watch for bare patches in your lawn.** Skunks may dig up pieces of sod to place in and around their own dens, preferring the softness of grasses to the chilly, potentially rocky earth in their hole. If bare spots and recent excavation appear, you may have a skunk in residence.
- **Watch out for holes.** Skunks may dig several holes for dens before choosing the one it will use to raise young. If you see a number of new holes in your gardens or around and under structures, you may have a skunk looking for a place to call its own. Once it chooses a den, the skunk may move from one hole to another

occasionally for reasons of its own. Before you simply fill in a random hole in your garden, pile some loose leaves in front of it and check in a day or so to see if the leaves have been disturbed, signaling that something may have come into or out of the hole. This is one way to be sure that you're not accidentally burying baby skunks alive, something that I'm sure readers of this book would not want to do.

- **It may smell**. Skunks generally don't spray in their own dens, but you may find yourself trying to track down the source of other kinds of animal smells: urine, feces, and the remains of a birth. These odors are all signs that something—maybe a skunk, maybe something else—has made a nest in your yard.

In the unlikely event that a skunk sprays you, you may experience some short-term discomfort, especially if the spray gets in your eyes. The National Capital Poison Center recommends these steps:

- **Eyes**. If your eyes are stinging, burning, or tearing, rinse them gently with cold water for up to fifteen minutes, or until the discomfort is gone. If the discomfort persists after that, seek medical assistance.
- **Lungs**. Breathing in the spray can irritate your lungs, and in rare cases it can aggravate asthma symptoms. Your inhaler can help you regain your normal breathing; if it doesn't help or if coughing persists for more than a few minutes, call your doctor for advice.
- **Vomiting**. In some individuals, the stench of skunk musk can induce vomiting. This discomfort usually passes quickly, but you will feel better once you begin to neutralize the odor. Hydrate with water or whatever fluids you can stomach.
- **Kill the smell**. Science has long since proved that tomato juice doesn't neutralize skunk odor, so don't bother with that very messy, very ineffective option. Here is what does work, according to poison control experts across the country:
  - one quart of household hydrogen peroxide (a standard bottle)

- one-quarter cup of baking soda
- one teaspoon of liquid dish detergent (preferably Dawn)

Mix this quickly in an open container and use it immediately. Apply to your own skin and clothing, your dog's fur, or to anything else that took the spray (but *not* to eyes or mouth). The dish detergent cuts the oil, just as it does for a dirty pot or an oil-saturated duckling. Wait five minutes and rinse, then repeat until the odor is gone.

If your eyes are burning or stinging, rinse them gently with cool water for fifteen minutes or until the discomfort stops. For skunk spray in your mouth, rinse with water and spit until the taste and odor are gone.

To neutralize the odor on objects like shoes or a deck floor, combine one cup of bleach with a gallon of water and wash with a sponge mop, or use rubber gloves and a handheld sponge.

If your clothing is saturated, first consider just putting them in a garbage bag and throwing it away. For any clothing you need to keep, wash the items with a laundry detergent made for heavy soils, or with a time-honored, old-fashioned cleanser called 20 Mule Team Borax, which is available at many hardware stores.

A number of commercial products are also available that make skunk odor a specific target. Try searching the web on "skunk odor killing products" to see the selection. If you live in an area with a lot of skunk activity, it may be worth your while to keep one of these in your home for emergencies—but hydrogen peroxide is cheap and easy to store, and chances are you already have dish detergent and baking soda in your kitchen.

With the potential for an encounter with a rabid skunk, it's time to take precautions to protect your pets against infection. Dogs are at particular risk, as they do not recognize the signs that a skunk is behaving erratically.

- **Vaccinate your dogs and cats for rabies**. If you have ferrets that go outside, vaccinate them as well.
- **Keep your cats and ferrets indoors.** Pets that are allowed to wander outdoors on their own recognizance are much more likely to run into trouble with wild animals.
- **When you walk your dog, keep it on a leash and maintain control of where it goes.** It only takes a second for a skunk under a bush to bite your sniffing dog's nose or face.
- **Contact animal control or law enforcement** if you see a skunk behaving aggressively, as this is not normal behavior for them.
- **Don't wait.** If your pet encounters a skunk or is bitten by one, contact your veterinarian or animal hospital immediately.

A Final Word

I have endeavored throughout this book to show the benefits and even the advantages of sharing our neighborhoods with wildlife, even if it takes some special effort on our part to protect parts of our property, gardens, and belongings from the most intrusive and obstreperous animals.

The fact is that humans' growing need for more housing, more places of business, and more manufacturing space will continue to encroach on the habitat that remains for animals, even though they beat us here by thousands or even millions of years. It is not fair to believe that we are somehow more entitled to this ground than they are, nor is it healthy for humans to dominate every square inch of the land on Earth. The need for balance between humans and their environment has never been more dire. Every animal that attempts to share our space with us brings us a message: They have run out of options. They must now live in harmony with the people around them if they hope to survive.

Whether it's a bear at your bird feeder, a fox family under your shed, a deer in your garden, or squirrels burying nuts in your flowerpots, we share the world with these furry and feathered neighbors, and they make our lives richer for the experience. When we find ways to live in harmony, everyone benefits. Please join me in loving our animal neighbors, and in helping to make a place for them in the world we all share.

Acknowledgments

In the year I spent researching this book, the reluctance of some wildlife specialists to talk to me about their work took me completely by surprise. Luckily, some people were kind and forthcoming, recognizing that I wanted only to help them tell the stories of the animals that cohabit with us in our neighborhoods, and how we can all live in the same environment without conflict.

I am indebted to homeowners David Oppenheimer in Asheville, North Carolina; Luane Haggerty in upstate New York; and Amy Irwin here in Rochester for their willingness to share their experiences with the wildlife in their own backyards. My old friend Michael Coombs sent me news stories about the deer in Tega Cay, South Carolina, which led to the perfect story for that chapter, as well as to my email interview with city manager Charles Funderburk. I thank Mike for his thoughtfulness, and Charlie for his time.

Wildlife professionals Linda Masterson of BearWise, North Carolina special projects biologist Ashley Hobbs, Richard Heilbrun, state Wildlife Diversity Program director for the Texas Parks and Wildlife Department, and Christopher Nagy, wildlife biologist at Mianus River Gorge in Westchester County, New York, and the cofounder of the Gotham Coyote Project, were all very generous with their time and insights about the challenges of managing animals and people when they come together in urban settings.

I am so pleased to be working once again with Rick Rinehart, my editor at Lyons Press, and with the talented team at Globe Pequot who always turn out such beautiful books. As ever, my agent Regina Ryan keeps my author career on track, and she provided the information that connected me with Chris Nagy and the Gotham Coyote Project, which proved extremely helpful in understanding how coyotes have made a home for themselves in the Bronx, of all places.

To so many friends—in particular, Ken, Rose-Anne, Kevin, Lisa, Martin, Bruce, Martha, Peter, Ruth, John, Georgia, Mark, Margery, and all of the other friends who take an interest in my writing and my endless stories about what I've learned—I truly could not stay motivated to do this without your support and encouragement.

And to my husband Nic, without whom there would be no books at all . . . I continue to be the luckiest woman in America. Let's go see more of it.

Selected Bibliography

Introduction
Chalasani, Radhika. "Photos: Wildlife Roams During the Coronavirus Pandemic." ABC News, April 22, 2020. Accessed December 10, 2024. https://abcnews.go.com/International/photos-wildlife-roams-planets-human-population-isolates/story?id=70213431.

Chapter 1: Black Bears
BBC Studios. "The Man Who Feeds Wild Black Bears." YouTube video. Posted December 5, 2008, by Bear Crimes, BBC Studios. Accessed July 11, 2024. https://www.youtube.com/watch?v=RB9uzMjiYSQ.
BearWise. Last accessed June 26, 2024. https://bearwise.org.
Cockcroft, Jessica. "Bear Attack Statistics." BearVault, August 26, 2023. Accessed July 3, 2024. https://bearvault.com/bear-attack-statistics.
Derworlz, Colette. "Research Shows Dogs Can Prompt Bears to Attack." *Calgary Herald*, June 9, 2014. Accessed July 12, 2024. https://calgaryherald.com/news/local-news/research-shows-dogs-can-prompt-bears-to-attack.
Dymburt, Andrew. "'I Thought I Was Going to Die': Pennsylvania Woman Survived Bear Attack, Recalls Terrifying Moments." ABC 7 Chicago, March 7, 2024. Accessed July 12, 2024. https://abc7chicago.com/woman-survives-bear-attack-pittsburgh-pennsylvania-black-pa/14499043/.
Get Bear Smart. Last accessed June 25, 2024. https://www.bearsmart.com/.
Herrero, Stephen. *Bear Attacks: Their Causes and Avoidance*, third edition. Lanham, MD: Lyons Press, Rowman & Littlefield Publishing Group, 2018.
Louisiana Wildlife and Fisheries Foundation. "Louisiana Black Bear Program." Accessed June 26, 2024. http://www.lawff.org/louisiana-black-bear.
Masterson, Linda. *Living With Bears Handbook*, expanded second edition. Costa Rica: PixyJack Press, 2021.
———. Personal interview with the author. June 25, 2024.
Minnick, Edith E. "Heroic Pomeranian, Smokie, Saves Owner from Bear Attack in Pennsylvania, Captivates Community with Bravery." *DeepHearting*, April 10, 2024. Accessed July 12, 2024. https://deephearting.com/heroic-pomeranian-smokie-saves-owner-from-bear-attack-in-pennsylvania-captivates-community-with-bravery.

New York State Black Bear Response Manual, third edition. New York State Department of Environmental Conservation, July 26, 2011. Accessed July 10, 2024. https://extapps.dec.ny.gov/docs/wildlife_pdf/bearsopm.pdf.

Rogers, Lynn. "How Dangerous Are Black Bears?" North American Bear Center. Accessed July 6, 2024. https://bear.org/bear-facts/how-dangerous-are-black-bears/.

Thompson, Taylor. "Feeding Bears Can Lead to Property Damage, Safety Issues, Fines, NC Wildlife Officials Say." WLOS-TV, May 1, 2024. Accessed July 11, 2024. https://wlos.com/news/local/feeding-bears-can-lead-to-property-damage-safety-issues-fines-nc-wildlife-officials-asheville-buncombe-county-ordinances.

TyAmongAnimals. "Let's Give the Bears Some Treats!" YouTube short. Posted April 23, 2023, by @TyAmongAnimals. Accessed July 11, 2024. https://youtube.com/shorts/eubIdUmsrLc?si=7fSTesDgs_251fBb.

WATE 6 On Your Side. "Don't Feed the Bears' Video Shows Gatlinburg Visitors Feeding Bears." YouTube video. Posted July 1, 2024, by WATE 6, Knoxville, TN. Accessed July 11, 2024. https://youtu.be/eBL_KUAbpYI?si=XNFz02bqScIETWVAb.

Weyler, Rex. "A Brief History of Environmentalism." Greenpeace, January 5, 2018. Accessed June 26, 2024. https://www.greenpeace.org/international/story/11658/a-brief-history-of-environmentalism/.

CHAPTER 2: MOUNTAIN LIONS

Applebome, Peter. "South Dakota Was Home to Mountain Lion Killed in Connecticut." *The Day* (New London, CT), July 27, 2011, 6. Accessed August 27, 2024. https://www.newspapers.com/image/972750002.

Benson, Judy. "Authorities Test Mountain Lion Hit by Car." *The Day* (New London, CT), June 14, 2011, 11. Accessed August 27, 2024. https://www.newspapers.com/image/972747655.

Bierman, Paul. "Clearcutting and Erosion in New England—the Photographic and Stratigraphic Record." *Vignettes*, July 3, 2008. Accessed August 30, 2024. https://serc.carleton.edu/vignettes/collection/24682.html.

Dockery, Stephen. "Mountain Lion Warning Issued for Greenwich." *The Day* (New London, CT), August 12, 2011, 22. Accessed August 27, 2024. https://www.newspapers.com/image/972764673.

French, Tom. "Mountain Lions in Massachusetts: Distinguishing Fiction from the Facts." Mass.gov. Accessed August 27, 2024. https://www.mass.gov/doc/mountain-lions-in-massachusetts-distinguishing-fiction-from-the-facts/download.

Loh, Tim. "State DEP Cops Deal with Cougars, Bears." *Connecticut Post*, August 28, 2011, C2. Accessed in the Record-Journal, Meriden, CT, August 27, 2024. https://www.newspapers.com/image/674321269.

Minetor, Randi. *Death in Rocky Mountain National Park: Accidents and Foolhardiness on the Continental Divide*. Guilford, CT: Lyons Press, 2020.

Mountain Lion Foundation. "Stay Safe." Accessed September 3, 2024. https://mountainlion.org/stay-safe/.

———. "A Timeline of Mountain Lions in the United States.". Accessed August 30, 2024. https://mountainlion.org/us/united-states/#timeline.

National Park Service. "Your Safety in Mountain Lion Habitat." Point Reyes National Seashore. Accessed Sept. 3, 2024. https://www.nps.gov/pore/planyourvisit/your safety_mountainlions.htm.

Tougias, Robert. *The Quest for the Eastern Cougar: Extinction or Survival?* Bloomington, IN: iUniverse, 2011.

US Fish and Wildlife Service, Northeast Region. "US Fish and Wildlife Service Concludes Eastern Cougar Extinct." News release, March 2, 2011. Accessed August 30, 2024. https://web.archive.org/web/20200728224833/https://www.fws.gov/northeast/ECougar/newsreleasefinal.html.

Velsey, Kim. "2 More Mountain Lion Sightings." *Hartford Courant*, June 13, 2011, B01. Accessed August 27, 2024. https://www.newspapers.com/image/247081322.

———. "A Big Pet Cat, But Whose?" *Hartford Courant*, June 14, 2011, A01. Accessed August 27, 2024. https://www.newspapers.com/image/247081707.

CHAPTER 3: BOBCATS

Beck, Abigail. "Bobcat Attacks Man in Arizona National Park. Here's What to Know About Bobcats." *Arizona Republic*, January 25, 2024. Accessed September 6, 2024. https://www.azcentral.com/story/news/local/arizona/2024/01/25/are-bobcats-dangerous/72338155007/#.

California Department of Fish and Wildlife. "Bobcat." Accessed September 6, 2024. https://wildlife.ca.gov/Conservation/Mammals/Bobcat.

———. "Living with Wildlife: Bobcat." Accessed September 6, 2024. https://nrm.dfg.ca.gov/FileHandler.ashx?DocumentID=202579&inline.

Christy, Matt. "Holcomb Signs Bobcat Hunting Bill into Law; DNR Must Establish Hunting Season by 2025." Fox 59, March 11, 2024. Accessed September 6, 2024. https://fox59.com/indiana-news/holcomb-signs-bobcat-hunting-bill-into-law-dnr-must-establish-hunting-season-by-2025/.

Connecticut Department of Energy and Environmental Protection. "Bobcat." Accessed September 6, 2024. https://portal.ct.gov/deep/wildlife/fact-sheets/bobcat.

Fowles, Gretchen. "Bobcat, *Lynx Rufus*." New Jersey Department of Environmental Protection. Accessed September 6, 2024. https://dep.nj.gov/wp-content/uploads/njfw/bobcat-fact-sheet.pdf.

Gelber, Ben. "Bobcats Returning to Parts of the Ohio Valley, a Century After Disappearing." WTRF.com, December 8, 2023. Accessed September 6, 2024. https://www.wtrf.com/ohio-valley/bobcats-returning-to-parts-of-the-ohio-valley-a-century-after-disappearing/.

Heilbrun, Richard. Personal interview with author. August 30, 2024.

Heilbrun, Richard, Julie Young, Derek Broman, and Terry Blankenship. "An Evaluation of Urban Bobcat Diet in the Dallas-Fort Worth Metropolitan Area." Unpublished data.

People for the Ethical Treatment of Animals. "The Jaws of Death: How Steel-Jaw Traps Torture and Kill Animals." PETA. Accessed September 6, 2024. https://www.peta.org/features/steel-jaw-trap-fur-cruelty/.

Young, Julie K., Julie M. Golla, Derek Broman, Terry Blankenship, and Richard Heilbrun. "Estimating Density of an Elusive Carnivore in Urban Areas: Use of Spatially Explicit Capture-Recapture Models for City-Dwelling Bobcats." *Urban Ecosystems*, February 8, 2019. Accessed September 5, 2024. https://doi.org/10.1007/s11252-019-0834-6.

CHAPTER 4: WILD TURKEYS

And-Hof Animals. "And-Hof Animals, Sanctuary for Farm Animals and Permaculture Is Different Than Other Animal Sanctuaries." Accessed September 24, 2024. https://www.and-hof-animals.org/family-of-animals.

Bascombe, Erik. "Turkey Task Force: Increasing Fowl Play Leaves Staten Island Scrambling for Answers." From the Scene podcast, *Staten Island Advance*, May 4, 2022. Accessed September 25, 2024. https://www.facebook.com/statenislandadvance/videos/turkey-task-force-increasing-fowl-play-leaves-staten-island-scrambling-for-answe/273202178158629.

———. "NY Officials: Turkeys Are Now 'Status Quo' and Need to Be 'Accepted' by Staten Islanders." *SILive*, February 26, 2024. Accessed September 24, 2024. https://www.silive.com/news/2024/02/ny-officials-turkeys-are-now-status-quo-and-need-to-be-accepted-by-staten-islanders.html.

Binkley, Collin. "Birds Gone Wild: Resurgent Turkeys Spar with Human Neighbors." Associated Press, October 13, 2017. Accessed September 25, 2024. https://apnews.com/article/d6a8c9df55fc49b28fa5d5a5108ef507.

BirdCentralPark (@BirdCentralPark). "The Day Ended Well . . ." *Manhattan Bird Alert*, posted May 7, 2024. Accessed September 24, 2024. https://x.com/BirdCentralPark.

Cristantiello, Ross. "Why Are There So Many Wild Turkeys Near Boston?" *Boston.com*, May 8, 2023. Accessed September 25, 2024. https://www.boston.com/news/wickedpedia/2023/05/08/why-wild-turkeys-boston-massachusetts/

Dalton, Kristin F. "Tackling Turkeys, a Growing Neighborhood Nuisance." *SILive*, April 16, 2018. Accessed September 24, 2024. https://www.silive.com/news/2018/04/neighborhood_nuisance_turkeys.html.

Editorial Board. "The Turkey Dilemma." *Staten Island Advance*, August 15, 2013. Accessed September 26, 2024. https://www.newspapers.com/image/1112563909/.

Gorman, Jessica Jones. "Angry Birds! Video Shows a Flock of Wild Turkeys Brawling on Staten Island." *SILive*, July 1, 2023. Accessed September 25, 2024. https://www.silive.com/news/2023/07/angry-birds-video-shows-a-flock-of-wild-turkeys-brawling-on-staten-island.html.

Hagley Vault. "The Bald Eagle Became America's National Symbol on June 20, 1782." Hagley Museum and Library, Wilmington, DE, December 10, 2020. Accessed September 26, 2024. https://www.hagley.org/research/news/hagley-vault/bald-eagle-became-americas-national-symbol-june-20.

Hargrave, Alex. "Talking Turkey: City of Buffalo Considers Ordinance That Would Ban Feeding Wild Turkeys." *Buffalo Bulletin*, January 11, 2023. Accessed September 25, 2024. https://www.buffalobulletin.com/news/article_4d3a2240-9131-11ed-89aa-bfb9539204c8.html.

Selected Bibliography

Kdonohue. "Talking Turkey." *Annotation*, blog of the National Historical Publications and Record Commission, National Archives. November 25, 2020. Accessed September 26, 2024. https://annotation.blogs.archives.gov/2020/11/25/talking-turkey.

Kennamer, James Earl. "What's Ailing Wild Turkeys?" National Wild Turkey Foundation, August 17, 2021. Accessed September 26, 2024. https://www.nwtf.org/content-hub/whats-ailing-wild-turkeys.

Miller, James E. "Wild Turkeys." US Department of Agriculture, Animal and Plant Inspection Services, Wildlife Damage Management Technical Series. January 2018. Accessed September 26, 2024. https://www.aphis.usda.gov/sites/default/files/Wild-Turkeys-WDM-Technical-Series.pdf.

Scalese, Roberto, Tiziana Dearing, and Walter Wuthmann. "Wild Turkeys in Brookline May Be Intimidating, but They Used to Almost Be Extinct." WBUR, November 25, 2019. Accessed September 25, 2024. https://www.wbur.org/radioboston/2019/11/25/turkey-conservation-rebound-urban.

Sherry, Virginia. "'Milestone Agreement' Reached to Address Staten Island Turkey Problem." *SILive*, March 7, 2016. Accessed September 24, 2024. https://www.silive.com/eastshore/2016/03/staten_island_turkey_agreement.html.

Sherry, Virginia N. "Fine-Feathered Friends Rally for Turkeys." *Staten Island Advance*, August 15, 2013, A1/A4. Accessed September 26, 2024. https://www.newspapers.com/image/1112563878/.

———. "Island Turkeys Safe and Sound in Saugerties." *Staten Island Advance*, September 26, 2013. Accessed September 26, 2024. https://www.newspapers.com/image/1112927104.

———. "Turkeys Getting It in the Neck." *Staten Island Advance*, August 13, 2013, A1/A12. Accessed September 26, 2024. https://www.newspapers.com/image/1112563617/.

Staten Islander News Service Staff. "Wild Turkey Control Is Not Necessary: They Have Been Here Longer Than Humans, Control Pest Insect Populations, and Do No Harm Aside from a Little Inconvenience." *Staten Islander*, February 10, 2023. Accessed September 25, 2024. https://statenislander.org/2023/02/10/wild-turkey-control-is-not-necessary-they-have-been-here-longer-than-humans-control-pest-insect-populations-and-do-no-harm-aside-from-a-little-inconvenience/.

"Turkey Fight!" YouTube video. Origin unknown. Posted by *Staten Island Advance*, July 1, 2023. Accessed September 25, 2024. https://youtu.be/AookE_LStL8.

US Department of Agriculture. Chapter XVI: "The Use of Egg Addling in Wildlife Damage Management. Human Health and Ecological Risk Assessment for the Use of Wildlife Damage Management Methods by USDA-APHIS-Wildlife Services." August 2018; peer-reviewed final October 2022. Accessed September 26, 2024. https://www.aphis.usda.gov/sites/default/files/16-egg-addling.pdf.

Chapter 5: Deer

Bowyer, Caroline. "Tega Cay Looking to Sterilize up to Two Hundred Deer." *Queen City News*, August 20, 2024. Accessed November 8, 2024. https://www.qcnews.com/news/u-s/york-county/tega-cay-looking-to-sterilize-up-to-200-deer/.

"Deer Resistant Native Plants." DirectNativePlants.com. Accessed November 4, 2024. https://directnativeplants.com/deer-resistant-native-plants.

Dys, Andrew. "Sharpshooters Take Aim: Tega Cay Doubles the Number of Deer to Kill, Will Donate Meat." *Rock Hill Herald*, January 23, 2024. Accessed November 8, 2024. https://www.heraldonline.com/news/politics-government/article284548225.html.

Funderburk, Charlie. Personal interview with author by email, November 6, 2024.

Hall, Ross. "When White-tailed Deer Become a Nuisance." Nova Scotia Department of Natural Resources and Renewables. Accessed November 4, 2024. https://novascotia.ca/natr/wildlife/nuisance/deer.asp.

Hanberry, Brice B. and Phillip Hanberry. "Regaining the History of Deer Populations and Densities in the Southeastern United States." *Wildlife Society Bulletin* 44, no. 3 (2020): 512–18. https://www.fs.usda.gov/rm/pubs_journals/2020/rmrs_2020_hanberry_b009.pdf.

Hewitt, David G. "Hunters and the Conservation and Management of White-Tailed Deer." *International Journal of Environmental Studies*, March 25, 2016. Accessed November 10, 2024. https://www.ckwri.tamuk.edu/news-events/hunters-and-conservation-and-management-white-tailed-deer.

Marks, John. "Tega Cay's Deer Quandary: Shoot, Sterilize, Move, or Let Them Be? Residents Sound Off." *Rock Hill Herald*, October 19, 2022. Accessed November 8, 2024. https://www.heraldonline.com/news/local/community/article267475753.html.

Ohio Canid Center. "Wolves in North America." Accessed November 9, 2024. http://www.ohiocanidcenter.com/wolves-in-north-america.html.

"Park Seed's Picks for Twenty-Five Best Deer-Resistant Plants for Your Garden." *Park Seed*. Accessed November 4, 2024. https://www.parkseed.com/blog/25-best-deer-resistant-plants-for-your-garden.

Shortreed, Nellie. "5 Things to Know: Deer Culling Begins in Tega Cay." *Queen City News*, January 27, 2024. Accessed November 8, 2024. https://qcnerve.com/deer-culling-tega-cay/.

White Buffalo. "Deer Management." Conserving Native Species and Ecosystems. Accessed November 8, 2024. https://www.whitebuffaloinc.org/deer-management.

Chapter 6: Bats

Arnold, Robert. "Texas Prison Officials Have New Plan for Warehouse Hosting Large Bat Colony." KPRC, October 31, 2024. Accessed November 19, 2024. https://www.click2houston.com/news/investigates/2024/10/31/bat-colony-forces-texas-prison-officials-to-rethink-demolition-plans/.

Barghouty, Leila. "They Thought They Found Their Dream Home—So Did Thousands of Bats." *Washington Post*, October 30, 2024. Accessed November 12, 2024. https://www.washingtonpost.com/climate-environment/2024/10/30/washington-bat-house-infestation/.

Bat Conservation International. "Bat Houses and Gardens: Guide to Gardening for Bats.". Accessed November 13, 2024. https://www.batcon.org/about-bats/bat-gardens-houses/.

———. "Huntsville Bat Colony." Accessed November 19, 2024. https://www.batcon.org/huntsvillebats/.

Bat Conservation Trust. "Bats as Pollinators." Accessed November 11, 2024. https://www.bats.org.uk/about-bats/why-bats-matter/bats-as-pollinators.

Government of Quebec. "Cleaning of an Environment Contaminated with Bat Droppings." Accessed November 12, 2024. https://www.quebec.ca/en/housing-territory/healthy-living-environment/cleaning-of-an-environment-contaminated-with-bat-droppings.

Gromicko, Nick, and Kenton Shepard. "Bat Infestation." International Association of Certified Home Inspectors. Accessed November 12, 2024. https://www.nachi.org/bat-infestation.htm.

Leibniz Institute for Zoo and Wildlife Research (IZW). "How an Urban Bat Differs from a Rural Bat." *ScienceDaily*, August 11, 2022. www.sciencedaily.com/releases/2022/08/220811142955.htm.

NYS Department of Environmental Conservation. "Bats of New York." Accessed November 11, 2024. https://extapps.dec.ny.gov/docs/administration_pdf/batsofny.pdf.

Rapp Learn, Joshua. "Cats Prey on Bats Around the World." The Wildlife Society, April 9, 2021. Accessed November 13, 2024. https://wildlife.org/cats-prey-on-bats-around-the-world/.

Rosenberg, Kenneth V., Adriaan M. Dokter, Peter J. Blancher, John R. Sauer, Adam C. Smith, Paul A. Smith, Jessica C. Stanton, Arvind Panjabi, Laura Helft, Michael Parr, and Peter P. Marra. "Decline of North American Avifauna." *Science*, September 2019. Accessed November 13, 2024. https://www.birds.cornell.edu/home/wp-content/uploads/2019/09/DECLINE-OF-NORTH-AMERICAN-AVIFAUNA-SCIENCE-2019.pdf.

Suh, Jonathan. "Winging It with New York Bats." *Hudson Valley Viewfinder*. Accessed November 11, 2024. https://www.scenichudson.org/viewfinder/winging-it-with-new-york-bats/.

US Department of the Interior. "Thirteen Awesome Facts About Bats." October 24, 2024. Accessed November 11, 2024. https://www.doi.gov/blog/13-facts-about-bats.

Visit Austin. "How to Experience Austin's Bats." Accessed November 12, 2024. https://www.austintexas.org/things-to-do/outdoors/bat-watching/.

Chapter 7: Coyotes

Colorado Division of Wildlife. "Most Commonly Asked Questions About Urban Coyotes." Accessed November 13, 2024. http://www.parkerpd.org/DocumentCenter/View/22686/Dept-of-Wildlife-Tips---MostCommonlyAskedQuestionsAboutUrbanCoyotes012010?bidId=.

Duncan, N., O. Asher, M. Weckel, C. Nagy, C. Henger, F. Yau, and L. Gormanzano. "Baseline Diet of an Urban Carnivore on an Expanding Range Front." *Journal of Urban Ecology* 6, no. 1 (2020): juaa021. Accessed November 18, 2024. https://academic.oup.com/jue/article/6/1/juaa021/5979496.

FOX 11 Los Angeles. "Video Shows Moment Coyote Attacks Toddler on Huntington Beach." YouTube video. Posted April 29, 2022. https://youtu.be/FNdJXqiVANc?feature=shared.

Gehrt, Stanley D., with Kerry Luft. *Coyotes Among Us: Secrets of the City's Top Predator.* Seattle, WA: Flashpoint Books, 2024.

Grant, Stan, Julie Young, and Seth Riley. "Assessment of Human-Coyote Conflicts: City and County of Broomfield, Colorado." USDA National Wildlife Research Center—Staff Publications, January 1, 2011. Accessed November 13, 2024. https://urbancoyoteresearch.com/sites/default/files/resources/Colorado%20Broomfield%20Report.pdf.

Henger, Carol S., Emily Hargous, Christopher M. Nagy, Mark Weckel, Claudia Wultsch, Konstantinos Krampis, Neil Duncan, Linda Gormezano, and Jason Munshi-South. "DNA Metabarcoding Reveals That Coyotes in New York City Consume Wide Variety of Native Prey Species and Human Food." *Peer Journal*, September 21, 2022. Accessed November 18, 2024. https://doi.org/10.7717/peerj.13788.

Inside Edition. "Dad Saves His Daughter from Coyote Attack." YouTube video. Posted December 6, 2022. https://youtu.be/65jkSc4kddY?feature=shared.

KATU News. "Dad, There's a Coyote! Girl Escapes Close Encounter in Backyard." YouTube video. Posted October 18, 2024. https://youtu.be/hPm2VxdU_WY?feature=shared.

Urban Coyote Research Project. "How to Avoid Conflicts with Coyotes.". Accessed November 13, 2024. https://urbancoyoteresearch.com/coyote-info/how-avoid-conflicts-coyotes.

———. "What Do Urban Coyotes Eat?" Accessed November 13, 2024. https://urbancoyoteresearch.com/faq/what-do-urban-coyotes-eat.

WYFF News 4. "Security Video Shows Man Wrestle with Coyote During Dog Attack." YouTube video. Posted January 15, 2024. https://youtu.be/8LlcwOvquH4?feature=shared.

Chapter 8: Moose

Alaska Department of Fish and Game. "What to Do About Aggressive Moose." Accessed November 21, 2024. https://www.adfg.alaska.gov/index.cfm?adfg=livewith.aggressivemoose.

Battinger, Brooke. "900-Pound Moose Roams into New Mexico City. See How Many Officers It Took to Lift It." *Sacramento Bee*, September 14, 2023. Accessed November 20, 2024. https://www.sacbee.com/news/nation-world/national/article279309149.html#storylink=cpy.

Cooper, Phil. "Moose in the Flower Garden, Oh My!" Found within *Moose in the Garden? It Happens.* Blog by Rich Landers. *The Spokesman-Review* (Spokane, WA), May 6, 2014. Accessed November 21, 2024. https://www.spokesman.com/blogs/outdoors/2014/may/06/moose-garden-it-happens/.

Goines, Ellen Miller. "New Moose Spotted Around Red River." *Santa Fe New Mexican*, August 19, 2024. Accessed November 20, 2024. https://www.santafenewmexican

Selected Bibliography

.com/news/local_news/new-moose-spotted-around-red-river/article_b6df8d5c-5e30-11ef-a303-93aa2023a771.html.

Grant, Bonnie L. "Types of Moose Deterrents—Tips on Keeping Moose Out of the Garden." *Gardening Know How*, February 15, 2023. Accessed November 21, 2024. https://www.gardeningknowhow.com/plant-problems/pests/animals/keeping-moose-out-of-garden.htm.

KOB 4. "Marty the Moose Relocated Again After Santa Fe Capture." YouTube video. Posted September 13, 2023. Accessed November 20, 2024. https://www.youtube.com/watch?v=AbymG2ePM9U.

Maine Department of Inland Fisheries and Wildlife. "Species Spotlights: Moose." Accessed November 20, 2024. https://www.maine.gov/ifw/fish-wildlife/wildlife/species-information/mammals/moose.html#diet.

Melchor, Laura Ojeda. "How to Keep Moose out of Your Garden and Orchard." *Gardener's Path*, July 15, 2020. Accessed November 21, 2024. https://gardenerspath.com/how-to/animals-and-wildlife/keep-moose-out/.

Michigan Department of Natural Resources "Michigan's Moose Lift." YouTube video. Posted April 2, 2013. Accessed November 22, 2024. https://youtu.be/jgs93C3wq6U?feature=shared.

Michigan Wildlife Council. "Magnificent Moose: The Comeback Story of Michigan's Upper Peninsula Herd." 2021. Accessed November 22, 2024. https://hereformioutdoors.org/stories/2019/10/09/magnificent-moose-the-comeback-story-of-michigans-upper-peninsula-herd/.

National Park Service. "Moose Safety." Noatak National Preserve, Alaska. Accessed November 21, 2024. https://www.nps.gov/noat/planyourvisit/moose-safety.htm.

———. "Species Spotlight: Moose." Accessed November 22, 2024. https://www.nps.gov/articles/species-spotlight-moose.htm.

National Wildlife Federation "Moose." NWF. Accessed November 22, 2024. https://www.nwf.org/Educational-Resources/Wildlife-Guide/Mammals/Moose.

Saskatchewan. "Brainworm or Moose Sickness." Accessed November 22, 2024. https://www.saskatchewan.ca/residents/environment-public-health-and-safety/wildlife-issues/fish-and-wildlife-diseases/brainworm-or-moose-sickness.

SUNY Environmental Science and Forestry. "Moose." SUNY ESF. Accessed November 22, 2024. https://www.esf.edu/aec/adks/mammals/moose.php.

Vanorio, Ame. "How to Safely Keep Moose Out of Your Garden." *Morning Chores*. Accessed November 21, 2024. https://morningchores.com/get-rid-of-moose/.

Vermont Fish and Wildlife. "Chapter 3: Moose." Vermont Fish and Wildlife. Accessed November 22, 2024. https://vtfishandwildlife.com/sites/fishandwildlife/files/documents/Learn%20More/Library/REPORTS%20AND%20DOCUMENTS/HUNTING/BIG%20GAME%20MANAGEMENT%20PLAN%20-%202010/5.%20MOOSE.pdf.

Wikipedia. "Dermacentor albipictus." Last updated August 31, 2024. Accessed November 22, 2024. https://en.wikipedia.org/wiki/Dermacentor_albipictus.

CHAPTER 9: SQUIRRELS

Baldwin, Marc. "What Controls the Caching Behaviour of Squirrels and How Do They Find Their Buried Nuts?" *Wildlife Online*. Accessed August 21, 2024. https://www.wildlifeonline.me.uk/questions/answer/what-controls-the-caching-behaviour-of-squirrels-and-how-do-they-find-their-bur.

Bryce, Emma. "Do Squirrels Remember Where They Buried Their Nuts?" *Scientific American*, November 20, 2023. Accessed August 21, 2024. https://www.scientificamerican.com/article/do-squirrels-remember-where-they-buried-their-nuts.

Golden Gate Bird Alliance. "Prevent Raptor Poisoning." Accessed August 26, 2024. https://goldengatebirdalliance.org/conservation/make-the-city-safe-for-wildlife/rodenticides-and-birds.

Holm, Jessica. "Daylight Robbery!" YouTube, video documentary (August 29, 1988). Posted on September 10, 2020, by FE3tMX5. https://youtu.be/jOzAfTBVtX8?feature=shared.

Mass Audubon. "A Campaign to Rescue Raptors." Accessed August 26, 2024. https://www.massaudubon.org/take-action/advocate/rescue-raptors.

McQuade, Denise B., Ernest H. Williams, and Howard B. Eichembaum. "Cues Used for Localizing Food by the Gray Squirrel (*Sciurus carolinensis*)." *Ethology* 72, no. 1 (1986): 22–30. https://onlinelibrary.wiley.com/doi/abs/10.1111/j.1439-0310.1986.tb00602.x

Minetor, Randi. *Backyard Birding and Butterfly Gardening*. Essex: CT: Lyons Press, 2021.

People for the Ethical Treatment of Animals. "Living in Harmony with Squirrels." PETA. Accessed August 21, 2024. https://www.peta.org/issues/wildlife/living-harmony-wildlife/squirrels.

Perky-Pet. "All About Squirrel Nests." Accessed August 21, 2024. https://www.perkypet.com/articles/squirrel-nests.

Town & Country Pest Solutions. "How to Get Rid of Squirrels and Keep Them from Coming Back." Rochester, NY. Accessed August 26, 2024. https://townandcountrysolutions.com/services/squirrels.

University of Toronto. "Animals Regulate Their Numbers by Own Population Density." *ScienceDaily*, December 7, 2000. www.sciencedaily.com/releases/2000/11/001128070536.htm.

CHAPTER 10: CANADA GEESE

All About Birds. "Canada Goose Life History." Cornell Lab of Ornithology. Accessed November 24, 2024. https://www.allaboutbirds.org/guide/Canada_Goose/lifehistory.

Blount, Sarah. "Waterfowl and Water Quality." National Environmental Education Foundation. Accessed November 25, 2024. https://www.neefusa.org/story/water/waterfowl-and-water-quality.

Department of Environmental Conservation. "Nuisance Canada Geese." New York State Department of Environmental Conservation. Accessed November 25, 2024. https://dec.ny.gov/nature/animals-fish-plants/nuisance-wildlife-species/canada-geese.

Department of Environmental Conservation. "When Geese Become a Problem." New York State Division of Fish, Wildlife, and Marine Resources. Accessed November 25, 2024. https://extapps.dec.ny.gov/docs/wildlife_pdf/geeseproblem.pdf.
Hanson, Harold. *The Giant Canada Goose*. Carbondale and Edwardsville, IL: Southern Illinois University Press, 1965. https://archive.org/details/giantcanadagoose00hans/page/n5/mode/2up.
Pearson, Erica. "How Was the Giant Canada Goose Rediscovered in Rochester After Being Declared Extinct?" *Minnesota Star Tribune*, November 16, 2024. Accessed November 24, 2024. https://www.startribune.com/rochester-giant-canada-goose-mayo-extinct/601181157.
Tikkanen, Amy. "US Airways Flight 1549." *Brittanica*, last updated March 26, 2025. Accessed November 25, 2024. https://www.britannica.com/topic/US-Airways-Flight-1549-incident.
Swan, Mike. "From Unwanted Game to Gourmet Dish: Cooking Canada Geese." Game and Wildlife Conservation Trust, May 1, 2024. Accessed November 24, 2024. https://www.gwct.org.uk/blogs/news/2024/january/from-unwanted-game-to-gourmet-dish-cooking-canada-geese.
Walking Mountains. "Curious Nature: Reintroduction of the Canada Goose." Walking Mountains Science Center, March 23, 2015. Accessed November 24, 2024. https://blog.walkingmountains.org/curious-nature/2015/03/reintroduction-of-the-canada-goose.
"Why Are 'Nuisance' Wild Birds Protected?" *Another Wild Goose Chase*, October 8, 2020. Accessed November 25, 2024. https://www.wildgoosechasers.com/bird-pigeon-control-geese-removal-blog/why-nuisance-birds-are-protected/.
Wildlife Services. "Management of Canada Goose Nesting." US Department of Agriculture Animal and Plant Health Inspection Service. Accessed November 25, 2024. https://www.aphis.usda.gov/sites/default/files/canada_goose.pdf.

Chapter 11: Raccoons

Blackwood, James. "Mobbed By Raccoons (25) Tuesday Night 03 Nov 2020." YouTube video. Posted November 3, 2020, by James Blackwood-Raccoon Whisperer. Accessed December 3, 2024. https://youtu.be/Ofp26_oc4CA?feature=shared.
Grider, Phillip. "'There Is a Beast They Call Aroughcun': Raccoons and Colonialism in Early America." *European Journal of American Studies* 19, no. 1 (2024). Accessed December 4, 2024. https://journals.openedition.org/ejas/21562.
Habitat Wildlife Center. "Four Steps to Follow If You See a Rabid Raccoon." Canfield, ON, Canada. Accessed December 3, 2024. https://habitatwildlifecontrol.ca/if-you-see-a-rabid-raccoon.
Humane Society of the United States. "Raccoons." Accessed December 3, 2024. https://www.humanesociety.org/animals/raccoons.
———. "What to Do About Raccoons." Accessed December 3, 2024. https://www.humanesociety.org/resources/what-do-about-raccoons.

May, Caitlin. "The Hidden Harm in Feeding Your Local Wildlife." US Fish and Wildlife Service. Accessed December 3, 2024. https://www.fws.gov/story/hidden-harm-feeding-your-local-wildlife.

Zap Termite and Pest Control. "Why Is There a Raccoon in My Backyard?" Sacramento, CA. Accessed November 26, 2024. https://www.zappest.com/about/our-blog/causes-raccoons-backyard.

CHAPTER 12: FOXES

Belford, Alan. "The Adaptable Red Fox." Lake Champlain Region, November 16, 2023. Accessed December 8, 2024. https://www.lakechamplainregion.com/story/2013/11/adaptable-red-fox.

Duda, Mark Damian, Martin Jones, Tom Beppler, Steven J. Bissell, Amanda Center, Andrea Criscione, Patrick Doherty, Gregory L. Hughes, Tristan Kirkman, Claudia Reilly, and Alison Lanier. "Trap Use, Furbearers Trapped, and Trapper Characteristics in the United States in 2015." Responsive Management National Office. Conducted for the Association of Fish and Wildlife Agencies, 2015. Accessed December 8, 2024. https://www.fishwildlife.org/application/files/3115/2106/4349/FINAL_AFWA_Trap_Use_Report_2015_ed_2016.pdf.

Historic Fort Snelling. "The Fur Trade." Accessed December 8, 2024. https://www.mnhs.org/fortsnelling/learn/fur-trade.

Mark J. Statham, Benjamin N. Sacks, Keith B. Aubry, John D. Perrine, and Samantha M. Wisely. "The Origin of Recently Established Red Fox Populations in the United States: Translocations or Natural Range Expansions?" *Journal of Mammalogy* 93, no. 1 (2012): 52–65. Accessed December 8, 2024. https://academic.oup.com/jmammal/article-abstract/93/1/52/899501?redirectedFrom=fulltext&login=false.

Pennsylvania Game Commission. "Wildlife Note: Foxes." PGC. Accessed December 8, 2024. https://www.pgc.pa.gov/Education/WildlifeNotesIndex/pages/foxes.aspx.

Resgetiloff, Kathy. "The Adaptable Red Fox Lives Among Us." *Bay Journal*, November 14, 2024. Accessed December 8, 2024. https://www.bayjournal.com/columns/bay_naturalist/the-adaptable-red-fox-lives-among-us/article_c98f0532-9057-11ef-b3d9-cb34a894bb78.html.

Wikipedia. "Fur Trade." Last updated December 22, 2024. Accessed December 8, 2024. https://en.wikipedia.org/wiki/Fur_trade#cite_note-42.

CHAPTER 13: ALLIGATORS

"Alligator Mauls Landscaper." *Miami Herald*, July 23, 2004, 3B. www.newspapers.com/image/651256735/?terms=%22Janie%20Melsek%22&match=1.

Almasy, Steve. "Disney Alligator Attack: Family of Dead Boy Won't Sue." CNN, July 21, 2016. Accessed December 19, 2024. https://www.cnn.com/2016/07/20/us/disney-alligator-attack/index.html.

Associated Press. "Final Report: Alligator Bit Boy's Head During Disney Attack." *Jacksonville.com*, August 22, 2016. Accessed November 23, 2024. https://www

Selected Bibliography

.jacksonville.com/story/news/2016/08/22/report-alligator-bit-boys-head-during-disney-attack/987192007/.

Cohen, Howard, and Martin Vassolo. "Body of Woman Attacked by Alligator Has Been Found." *Miami Herald*, June 10, 2018, 3. www.newspapers.com/image/654446712/?terms=%22Shizuka%20Matsuki%22&match=1.

"Death Spurs New Gator Policy." *Orlando Sentinel*, August 5, 2004. www.orlandosentinel.com/news/os-xpm-2004-08-05-0408050149-story.html.

Florida Fish and Wildlife Conservation Commission. "Alligator Bites on People in Florida." November 2021. Accessed November 23, 2024. https://www.documentcloud.org/documents/23686417-alligator-gatorbites.

———. "Human-Alligator Incidents Fact Sheet." https://myfwc.com/media/1776/human-alligatorincidentfactsheet.pdf. Accessed November 23, 2024.

HHMI Tangled Bank Studios. "Conservation Comeback: The American Alligator." Wild Hope, Jan. 10, 2024, accessed Nov. 23, 2024. https://www.wildhope.tv/article/the-american-alligator.

Kridel, Kristen. "Gator Kills Swimmer in Canal." *Ocala Star Banner*, Jul7 17, 2005. Accessed November 23, 2024. https://www.ocala.com/story/news/2005/07/17/gator-kills-swimmer-in-canal/31339397007/.

Lollar, Kevin. "Friends Remember Gator Victim as Talented Poet, Free Spirit." *Fort Myers News-Press*, September 28, 2004, 15. www.newspapers.com/image/219618888/?terms=%22Michelle%20Reeves%22%20alligator&match=1.

Louisiana Department of Wildlife and Fisheries. "Alligator Management." Accessed November 23, 2024. https://www.wlf.louisiana.gov/page/alligator-management.

McCloud, Cheryl. "Alligators Can Be Found in Every Florida County: What Are the Chances of Being Bitten?" *TC Palm*, February 20, 2023. Accessed November 23, 2024. https://www.tcpalm.com/story/news/local/2023/02/21/alligator-attack-florida-gator-kills-boy-disney-how-many-attacks-are-there-what-should-you-do-bitten/69926522007/.

"Public Service Set for Gator Victim." Fort Myers News-Press, July 29, 2004, 3B. www.newspapers.com/image/220623740/?terms=%22Janie%20Melsek%22&match=1.

Roberto Santiago, Diana Moskovitz, and Darran Simon, "Trappers Stalk a Killer Gator," *Miami Herald*, May 12, 2006, 1, 2. www.newspapers.com/image/654223871.

Stout, Byron. "Living with Alligators: Problems Grow with Population," *Fort Myers News-Press*, October 11, 2004, 1, 3. www.newspapers.com/image/220718060.

Chapter 14: Skunks

Centers for Disease Control. "Rabies in the United States: Protecting Public Health." CDC, June 21, 2024. Accessed December 10, 2024. https://www.cdc.gov/rabies/php/protecting-public-health/index.html.

Dragoo, J. W. "Skunk." Dragoo Institute for the Betterment of Skunks and Skunk Reputations. Accessed December 9, 2024. http://dragoo.org.

Ellison, Garret. "Is Grand Rapids trapped Under a 'Thick, Immovable Cloud' of Skunk Odor?" *MLive*, July 21, 2014. Accessed December 9, 2024. https://www.mlive.com/news/grand-rapids/2014/07/is_grand_rapids_trapped_under.html.

Hicks, Justin P. "Michigan Warns Residents of Possible Rabies Outbreak in Skunks." *MLive*, December 6, 2023. Accessed December 10, 2024. https://www.mlive.com/public-interest/2023/12/michigan-warns-residents-of-possible-rabies-outbreak-in-skunks.html.

Humane Society of the United States. "What to Do About Skunks." Accessed December 8, 2024. https://www.humanesociety.org/resources/what-do-about-skunks#encounter.

Indiana Department of Health. "Skunk Rabies Circulating in Southern Indiana." News release, February 9, 2024. Accessed December 9, 2024. https://events.in.gov/event/idoh_news_release_skunk_rabies_circulating_in.

Iowa Health and Human Services. "Rabies." HHS. Accessed December 10, 2024. https://hhs.iowa.gov/center-acute-disease-epidemiology/epi-manual/reportable-diseases/rabies.

Kemp, Christopher. "On the (Very Smelly) Trail of the Skunk Takeover." *Outside*, July 8, 2014. Accessed December 9, 2024. https://www.outsideonline.com/outdoor-adventure/environment/very-smelly-trail-skunk-takeover/.

Lawrence, Jp. "Minnesota Shows 'Significant Increase' in Rabies Cases, Driven by Rabid Skunks." *Minnesota Star-Tribune*, August 2, 2024. Accessed December 9, 2024. https://www.startribune.com/minnesota-shows-significant-increase-in-rabies-cases/600673659.

Ma, Xiaoyue, Cassandra Boutelle, Sarah Bonaparte, Lillian A. Orciari, Rene E. Dondori, Jordona D. Kirby, Richard B. Chipman, Christine Fehlner-Gardiner, Cin Thang, Veronica Gutiérrez Cedillo, Nidia Aréchiga-Ceballos, Yoshinori Nakazawa, and Ryan M. Wallace. "Rabies Surveillance in the United States During 2022." *AVMA Publications* 262, no. 11 (2024): 1518–25. Accessed December 10, 2024. https://avmajournals.avma.org/view/journals/javma/262/11/javma.24.05.0354.xml.

Mueller, Marcus. "Madison Wildlife Control: Signs of a Skunk Den." Skedaddle Humane Wildlife Control, December 3, 2022. Accessed December 9, 2024. https://www.skedaddlewildlife.com/location/madison/blog/signs-of-a-skunk-den/.

Poison Control. "What Happens If a Skunk Sprays Me?" National Capital Poison Center. Accessed December 8, 2024. https://www.poison.org/articles/what-happens-if-a-skunk-sprays-me-213.

Seton, Ernest Thompson. *Life-Histories of Northern Animals, Volume II: Flesh-Eaters*. New York: Charles Scribner & Sons, 1909. https://archive.org/details/lifehistoriesofn02seto/page/n9/mode/2up?view=theater.

US Department of Agriculture. Farmers' Bulletin No. 587. *Economic Value of North American Skunks*. June 1923. https://archive.org/details/CAT87203967/page/1/mode/2up?view=theater.

INDEX

Adirondack Park, New York, 2, 18
Alaska, 103, 113, 119, 149
alligators, *168*; bites and infection, 177; capturing, 170–71, 177; fatal attacks by, 169–71, 172, 173–74, 175–77; feeding, 178; Florida population of, 172; habitats, 171–72; hunted to near extinction, 174; hunting behaviors, 168, 171; odds of injury from, 172; pets and, 179; protections, 174–75; safety tips, 177–79; swimming safety and, 178
American black bear. *See* bears
American Ornithologists' Union (AOU), 136, 137
And-Hof Animals Sanctuary for Farm Animals + Permaculture, 54–55
Anholt, Jim, 176
Animal Damage Control Act (1931), 34
AOU. *See* American Ornithologists' Union
Asheville, North Carolina, 7–11, 19–20
Austin, Texas, 85
automobiles. *See* cars

Backyard Birding (Minetor), 35
bans. *See* legislation
bat houses, 89
bats, *78*; attracting, 88–89; benefits of, 84, 88; at Carlsbad Caverns National Park, 83; cats killing, 89; damage from, 87; deterring, 90–91; fear of, 79–80; guano-related illnesses from, 87–88; in houses, 78, 79–82, *82*, 87, 90–91; at Huntsville State Prison, Texas, 85–87; removing, humanely, 82–83, 86–87; signs of, 90; species and prevalence of, 83–84; on threatened species list, 83; urban adaptations of, 84–85; white nose syndrome decimating, 81
Bear Aware (program), 12
bear-resistant canisters, 15
bears, *6*; in Asheville, North Carolina, 7–11, 19–20; attacks compared to dog attacks, 16; attacks in presence of dogs, 24–25; bird feeders and, 9, 22–23; cars and, 17, 23–24; coexisting with and deterring, 20–27; cubs in backyards, *11*; diet, 8; escaping, 27; euthanization of, 18, 21, 25; fatal interactions with, 15; feeding, 18–20, 21; food conditioning issue for, 17–18; habituation, 16, 17; hiking safety and, 26–27; historical treatment of, 13–14; homeowners response to, 7; human behavior adapting to, 17–18, 20; mothers defending cubs, 15–16; population growth, 12–13, 15; pre-hibernation behavior, 8, 14; on threatened species list, 14; trap-and-relocate operations, 18; up-close interactions with, 9–11. *See also* grizzly bears
bear spray, 10, 26–27
BearVault, 15
BearWise (program), 8, 12
Beausoleil, Rich, 12

Beck, Tom, 12
Bierman, Paul, 33
bird baths, *125*
bird feeders: bears and, 9, 22–23; coyotes and, 105; foxes and, 159; squirrels and, 127–29; turkeys and, 62
black bear. *See* bears
bobcats, *40*, *43*; in Dallas-Fort Worth, Texas, 41–45; decline of, 46; diet, 44–45; farming impact on, 46; feeding, 42, 45, 47; habitat, 45; hiking safety and, 48–49; pets safety and, 42, 47–48; population and density of, 43–44, 45–46; protection efforts, 46–47; safety tips, 47–49; size, 41
Boston, Massachusetts, 58–59
bounty hunting, 151; legislation and, 34; livestock threat and, 2–3, 34; mountain lion, 33, 34
Buffalo, New York, 61

California: bobcat hunting ban in, 47; coyote attacks in, 98; grizzly bears historical treatment in, 13–14; Indigenous cultures on coyotes in, 101–2; mountain lion protections in, 35; SGAR ban in, 131
Canada geese, *132*; catching and releasing, 143; diet, 140–41; eating meat of, 134, 135; egg collection impact on, 135; extinction threat historically for, 132, 135–36; fecal matter of, 139; feeding, 140–41; giant, 137; goslings, 133–34, *138*; management and deterrents, 140–43; migration stopped by, 138–39; nest disturbance, 142–43; in New York, 133–34, 137–38; pond deterrents for, 141–42; protections for, 135–36, 139–40; in Rochester, Minnesota, 136–37; safety concerns with, 139; species, 136

Carlsbad Caverns National Park, New Mexico, 83
cars: bear-proofing, 23–24; bears and, 17, 23–24; deer and, 66; moose threat to, 112; turkey damage to, 55
Carson, Rachel, 14
cats (domestic): bats killed by, 89; bobcats compared with, 41–42, 49; coyotes threat to, 95, 98; skunk safety and, 193; turkey aggression towards, 59
cats, feral, 104
Catskill Animal Sanctuary, 58
Catskill Park, New York, 2, 18
cattle. *See* livestock
Chicago, Illinois: coyotes in, 1, 98–99, 101; Urban Coyote Research Project in, 98–99, 101, 104
children: alligator threat to, 173–74; coyote threat to, 98; mountain lions and, 35, 37
Chow, Pizza Ka Yee, 127
climate change, 73–74, 113, 115
Cockcroft, Jessica, 15
Colorado: Bear Aware program in, 12; bounty hunting in, 34; mountain lion attacks in, 35–36
compost bins: bear safety and, 24; coyotes and, 105; raccoons and, 152–53
Connecticut: bobcat protections in, 46; bounty hunting mountain lions in early, 33; mountain lions in, 29, 30–32
conservation movement: beginning of, 3; deer and, 73
controversy, 5
Cooper, Phil, 112
cougar. *See* mountain lions
COVID-19 pandemic, 1
coyotes, *92*, *96*; adaptation and expansion of, 102–3; attacks from, 97–98, 99; bird feeders and, 105; in cities during COVID-19 lockdown, 1; coexisting with, 105–7; deer

management by, 65, 66; diet, 94–95, 99; feeding, 105; fences for, 106; feral cats and, 104; help for removing, 104; Indigenous cultures on, 101–2; as pets threat, 65, 99, 106; population and range of, 92, 101; programs for protection of, 71; rodent control benefits of, 96–97, 100–101, 103–4; scat research, 93–94; social system of, 103; speed, 100; threat as minimal, 98–99
Coyotes Among Us (Gehrt and Luft), 98–99, 101, 103

Dallas-Fort Worth, Texas, 41–45
Dalton, Kristen, 55
Davis, Bo, 169–71
Daylight Robbery (documentary), 128
Death in Rocky Mountain National Park (Minetor), 123
Death in the Everglades (Minetor), 177
deer, *64, 71*; backyards as ideal for, 65; cars and, 66; climate change and, 73–74; coyotes eating, 95; deterrents, 74–77; diet, 65–66; differing opinions about, 4; feeding ban, 66; fences, 75–76; game preserves for, 73; hunting, historically, 72–73; hunting, in residential areas, 67–68, 74; moose habitat shared with, 114–15; mountain lions killing, 69–70; plants that deter, 76–77; population growth, 65, 67; predator reduction relation to number of, 69–71, 72; relocation, 66–67; sterilization program, 68–69; in Tega Cay, South Carolina, 65–69; tick-borne illnesses and, 74. *See also* moose
Department of Environmental Protection (DEP), 31, 32
deterrents. *See specific animals*
disease. *See* infection/infectious diseases
Disney Resort, Florida, 173–74

dogs: bear attacks compared with attacks by, 16; bears and safety for, 24–26; bobcat safety and, 47–48; coyotes threat to, 98, 99; food storage, 23; as geese deterrent, 142; guard, 39; moose safety and, 117; mountain lion safety and, 37; rabies safety for, 192–93; skunks spraying, 187; turkey safety and, 62

endangered species, 89; bats as, 83; bobcats as, 46, 47; mountain lions as, 29, 34
Endangered Species Act (1973), 3, 175
Endangered Species Conservation Act (1969), 14
Endangered Species Preservation Act (1966), 34
environmental legislation. *See* legislation
Estes Park, Colorado, 35–36
Esty, Daniel C., 31
euthanization, of bears, 18, 21, 25
extinction/near-extinction: geese and, 132, 135–36; of grizzly bears in California, 14; hunting to, 3, 52–53, 72, 114, 132, 174; mountain lions historical, 29–30

fear: of alligators, 176–77; alligators lack of, 174; of bats, 79–80; of bears, 16; bears lack of, 18, 21; of coyotes, 98; geese lack of, 140; moose, of humans, 119; mountain lions, of humans, 36; of skunks, 181–83; turkeys lack of, 52
feeding wildlife: alligators and, 178; bears and, 18–20, 21; bobcats and, 42, 45, 47; coyotes and, 105; geese and, 140–41; legislation on, 19, 60–61, 66; moose and, 119; raccoons and, 145–49; turkeys and, 60–61, 62; USFWS on dangers of, 147–48

fences: coyote deterrent, 106; deer, 75–76; fox deterrent, 167; maintaining, 48; moose, 118
Florida: alligator attacks in, 169–71, 173–74, 175–77; alligator population in, 172; bear population threat in, 14; mountain lion protections in, 34
Florida Fish and Wildlife Conservation Commission (FWC), 172, 174, 176, 177
food, pet. *See* pet food
food storage, bears and, 23
foxes, 1, *156*; backyard behaviors, 157–60, 164; bird feeders and, 159; coexisting with and deterring, 164–67; dens, 159, *163*, 166–67; diet of, 158, 160–61; fences for, 167; fur industry impact on, 161, 162–64; hunting and fur trade impact on, 162–64; pets and, 165; population and spread of, 160; reputation, 164; species and subspecies, 161–62
Funderburk, Charles, 65, 66–67, 68, 69
fur industry, 2–3; bobcats threat from, 46; foxes impacted by, 161, 162–64; modern, 151, 164; raccoons and, 149–51; skunks and, 185–86
FWC. *See* Florida Fish and Wildlife Conservation Commission

Galante, Lee Ann, 25
game preserves. *See* preserves and sanctuaries
garages: bears and, 17, 21; locking trash in, 21, 152, 165, 189, 190
garbage: bears and, 8–9, 21; locking-up/securing, 8–9, 21, 152, 165, 189, 190
garden plants/trees: deer deterrents and, 76–77; foxes destruction to, 158; geese and, 141; moose and, 113, 118–19; raccoons and, 153; skunks and, 186–87, 189–90; squirrels damage to, 125–26

Gatlinburg, Tennessee, 19
geese. *See* Canada geese
Gehrt, Stanley, 98–99, 101, 103
Gotham Coyote Project, 93–94, 97
grills, cleaning: bears and, 22; raccoons and, 154
grizzly bears: fatal attacks by, 16; food conditioning of, 17; historical treatment of, 13–14

habitat preservation. *See* preserves and sanctuaries
Haggerty, Luane and Peter, 157–59
Halfpenny, James, 190
Hansen, Tracy, 169–71
Hanson, Harold, 137
Heilbrun, Richard: on bats, 86; bobcats and, 42–43, 44–45
Henrietta, New York, 133–34
Herrero, Stephen, 16, 17, 24–25
hiking: bear safety while, 26–27; bobcat safety while, 48–49; mountain lion safety while, 37–38
Hobbs, Ashley, 19–20
Holm, Jessica, 128
housing development, 195; alligators and, 173; bear habitats and, 16–17; mountain lions impacted by, 33
Hristienko, Hank, 24
human behaviors: deer and, 65; wildlife adapting to, 3, 16–17, 42
human-wildlife relationship: balance in, 3, 57, 195; differing opinions on, 3–4; informed choices around, 5, 17–18, 46, 57, 106; questions to consider with, 99–100. *See also specific animals*
Hunter, Don, 35–36
hunting (by humans): alligators, 174; bobcat, 47; deer historically, 72–73; deer in residential areas, 67–68, 74; to extinction/near extinction, 3, 52–53, 72, 114, 132, 174; foxes,

Index

162–64; moose, 114; of mountain lions, 33–34; turkey, 52–53, 59. *See also* bounty hunting; fur industry
Huntsville State Prison, Texas, 85–87

Illinois. *See* Chicago
illness. *See* infection/infectious diseases; rabies
Indiana: bobcat hunting in, 47; rabid skunks in, 188
Indigenous cultures, 69; on coyotes, 101–2; fox hunting and, 163; moose eaten by, 113; raccoons and, 149; skunk trapping by, 185; turkeys and, 52
infection/infectious diseases: alligator bites and, 176, 178; from bats, 87–88; in bats, 81; moose, 115; turkeys and, 57–58. *See also* rabies
insect-control benefits, 185–86. *See also* ticks
International Association for Bear Research and Management, 12
International Human–Bear Conflicts Workshop, 12

Jean-Baptiste, Kelsey, 56

legislation: Animal Damage Control Act (1931), 34; bobcat, 46–47; on bobcat hunting in California, 47; bounty hunting, 34; endangered species, 3, 14, 34, 175; on feeding bears, 19; on feeding deer, 66; on feeding turkeys, 60–61; fur farming, 151; Migratory Bird Treaty Act (1918), 135–36, 139; on mountain lions in California, 35; on SGAR in California, 131
Life Histories of Northern Animals, Volume II (Seton), 183
livestock: bears and, 14; bobcat safety and, 48; bounty hunting and threats to, 2–3, 34; deer predators killed to protect, 70–71; food storage, 23; mountain lion safety for, 38–39; mountain lions threat to, 33, 34, 36; wildlife preying on, 2–3
Living With Bears Handbook (Masterson), 8, 9, 13
Los Angeles, California, 101
Louisiana: alligator protections in, 174–75; bear population threat and recovery in, 14; preserves and sanctuaries in, 14–15, 174–75; skunks, 184–85
Luft, Kerry, 98–99, 101, 103
lynx. *See* bobcats

Maine: deer in, 72; moose population in, 111; moose protections in, 114; mountain lions in, 29
Make Way for Ducklings (McCloskey), 133–34
Massachusetts: deer in, 72; moose in, 114; mountain lions in, 30, 33; turkeys in, 52, 58–59
Masterson, Linda, 8, 9, 12, 13, 15, 16–18, 25–26
May, Caitlin, 147–48
Mayo, Charles, 136
McCarter, John, 30
McCloskey, Robert, 133–34
Melsek, Janie, 175–77
Miedema, Mark, 35–36
Migratory Bird Treaty Act (1918), 135–36, 139
Minetor, Randi (books by), 35, 123, 177
Minnesota, 115; bears in, 15, 19; geese in, 136–37; mountain lion in, 31; skunks in, 188
Montana, 34; on feeding turkeys, 60; mountain lion research in, 31
moose, *108*, *112*; attacks from, 110; brainworms threat to, 115; breeding season threat from, 111; cars and, 112; climate change impacts on, 113, 116; deterrents, 118–19; diet,

111–12, 113, 115, 116; feeding, 119; fences for, 118; garden damage by and deterrents for, 113, 118–19; habitat, 109, 110, 114–15; hunting, 114; population historically, 113–14; protections for, 114; relocation of, 110–11; safety tips for encounters with, 116–18; in Santa Fe, New Mexico, 109–11; ticks and, 115; warm weather impact on, 115–16
motion-activated alarms: for bobcats, 48; for mountain lions, 39; for raccoons, 154
motion-activated sprinklers, 75
Mountain Lion Foundation, 33, 35, 36, 39
mountain lions, 1, *28*, *32*; bounty hunting of, 33, 34; California legislation protecting, 35; deer killed by, 69–70; 1800s foresting impact on, 32–33; extinction status historically, 29–30; fatalities from, 35–36; global distribution of, 36; habitat needs, 32; hiking safety and, 37–38; livestock safety around, 38–39; livestock threat from, 33, 34, 36; in New England, 29–32; other names for, 28, 30; poisoning, 71; range of, 31; risk of attack by, 36; safety tips, 36–39; stigma, 33; timeline of American, 33–34
Murray, Kevin Albert, 169–71

Nagy, Christopher, 93–94, 96–97, 100–101
National Wildlife Federation (NWF), 116
National Wild Turkey Federation (NWTF), 58
New England: mountain lion impact by foresting in 1880s, 33; mountain lion status and sightings in, 29–32; turkeys in, 52–53. *See also specific states*
New Mexico: bat viewing in, 83; moose in Santa Fe, 109–11

New York: bats dying from white nose syndrome in, 81; feeding turkeys in, 61; foxes in, 158–59; geese in, 133–34, 137–38; preservation areas in, 2, 18; squirrels in house in, 121–23; turkey population in, 51–53; turkeys historically in, 52–53
New York City, 53; coyotes in, 10, 93–97; Staten Island turkeys in, 51–52, 54–56, 58
New York State Department of Environmental Conservation (NYSDEC): on bears, 18; on geese, 138, 141, 142; on turkeys, 54–55, 56, 57, 58
North Carolina: bears in Asheville, 7–11, 19–20; Pisgah National Game Preserve in, 73
Northport, Florida, 169–71
Novo, Michael, 1
NWF. *See* National Wildlife Federation
NWTF. *See* National Wild Turkey Federation
NYSDEC. *See* New York State Department of Environmental Conservation

Ohio, 70, 187; bobcats in, 46–47; mountain lions in, 33
Oppenheimer, David, 7–11, 19

panther. *See* mountain lions
People for the Ethical Treatment of Animals (PETA): on deer relocation, 66–67; on squirrel damage to lawns, 126
pet doors, raccoon-proofing, 153–54
pet food: bears and, 8; dangers for wildlife, 148; location and storage of, 23, 47, 153, 189
pets: alligators and, 179; bobcats and, 42, 47–48; coyotes threat to, 65, 99, 106; foxes and safety for, 165;

turkey safety and, 62; turkeys as, 51. *See also* cats; dogs
Pisgah National Game Preserve, North Carolina, 73
plants, garden. *See* garden plants/trees
poisoning: coyotes threat from, 96; mountain lions, 71; squirrels, 130–31
preserves and sanctuaries: for alligators, 175; And-Hof Animals Sanctuary for Farm Animals + Permaculture, 54–55; Catskill Animal Sanctuary, 58; for deer, 73; in Louisiana, 14–15; in New York, 2, 18; in North Carolina, 73
property damage: from bears, 19; from turkeys, 55. *See also* garden plants/trees
puma. *See* mountain lions

rabies: in bobcats, 48; dog safety from, 192–93; human death from, 188–89; in raccoons, 155; in skunks, 187–89
raccoons, 1, *144*; adaptive behaviors, 151–52; coexisting with and deterring, 152–54; coyotes killing/eating, 94; evolution, 149; feeding, 145–49; fur industry history and, 149–51; pet doors secured against, 153–54; rabies in, 155; urban habitats, 148, *148*
rats. *See* rodent control, coyotes and; rodent poisons
recycling containers, 21
reforestation efforts, 34–35
research, 5, 12; coyote, 93–94, 98–99, 101, 104; mountain lion, 31
Ring doorbell camera, 9–10, 19
Rochester, Minnesota, 136–37
Rochester, New York, 121–23
Rocky Mountain National Park, 35–36
rodent control, coyotes and, 96–97, 100–101, 103–4

rodent poisons: coyote threat from, 96; SGAR, 130–31
Rogers, Lynn, 15–16

sanctuaries, wildlife. *See* preserves and sanctuaries
Sanibel Island, Florida, 175–77
Santa Fe, New Mexico, 109–11
Scarpitti, David, 59, 60
Scats and Tracks of North America (Halfpenny), 190
SCDNR. *See* South Carolina Department of Natural Resources
Seattle, Washington, 101
second generation anticoagulant rodenticide (SGAR), 130–31
SECR. *See* spatially explicit capture-recapture
Seton, Ernest Thompson, 183
SGAR. *See* second generation anticoagulant rodenticide
Silent Spring (Carson), 14
SIUH. *See* Staten Island University Hospital
skunks, *180*; coexisting with and deterring, 189–92; dens, *186*, 190–91; diet, 185–86; fear of, 181–83; fur industry and, 185–86; insect-control benefits of, 185–86; rabies in, 187–89; species and spread of, 184–85; spraying triggers and behavior, 180, 183–84, 187; spray odor remedies, 191–92; urban habitats, 187
South Carolina Department of Natural Resources (SCDNR), 66
South Dakota, 31, 33, 35
spatially explicit capture-recapture (SECR), 44
squirrels, *120*, *126*; bird feeders and, 127–29; caches/caching behavior of, 126–28; deterrents, 128–30; diet, 125–26; habitat and nests, 124–25, 126, 127; in houses, 121–23, 130–

31; intelligence, 127–28; poisoning, 130–31; reproduction cycle, 123–24; species of, 124; trapping and relocating, 122–24
Staten Island, New York City, 51–52, 54–56, 58
Staten Island University Hospital (SIUH), 51, 54
Sullenberger, Chesley "Sully," 139

TDCJ. *See* Texas Department of Criminal Justice
Tega Cay, South Carolina, 65–69
Tennessee, 19
Texas: bats in, 85–87; bobcats in Dallas-Fort Worth, 41–45
Texas Department of Criminal Justice (TDCJ), 86–87
Texas Parks and Wildlife Department, 42, 86
threatened species list, 89; bats on, 83; bears on, 14. *See also* endangered species
ticks: deer and illness from, 74; moose and, 115; turkeys controlling, 55
tracks, identifying, 190
trash. *See* garbage
trees, garden. *See* garden plants/trees
turkeys, 50, 53; bird feeders and, 62; breeding, 50; car damage by, 55; diseases and parasites of, 57–58; eating, 57–58; eggs addling (destruction), 61; feeding, 60–61, 62; fights, 55–56; historically in New York, 52–53; in Massachusetts, 52, 58–59; pecking order and aggression of, 59–60; as pets, 51; pets and, 62; population, 51–53, 60; property damage from, 55; relocation of, 54–55, 58–59; safety when encounters with, 60, 62; on Staten Island, New York, 51–52, 54–56, 58; tick control by, 55; tips for coexisting with, 61–63; trapping and killing, 56–57

Urban Coyote Research Project, 98–99, 101, 104
urban development, 195; coyotes and, 102–3; deer population and, 73; wildlife and, 2. *See also* housing development
US Department of Agriculture (USDA): deer hunting by, 67–68; geese deterrents and, 143; turkey programs, 54–55, 56–57
US Fish and Wildlife Service (USFWS), 14, 29–30; on feeding wildlife dangers, 147–48; on geese deterrents, 143; mountain lions hunted by, 33–34
Utah, 43; bounty programs in, 34; mountain lions hunted in, 33–34

Vanderbilt, George Washington, 73
Vermont, 114
volunteering, 12

Ward, Steve, 30
Washington, 112; bats in, 87; bear research in, 12; bounty hunting in, 34; coyotes in Seattle, 101
White Buffalo (organization), 68–69
white nose syndrome, for bats, 81
Wildlife NYC, 53
wildlife preserves and sanctuaries. *See* preserves and sanctuaries
wild turkeys. *See* turkeys
wolves: deer numbers relation to, 69; eradication of, 71; as livestock threat, 70 71

Young, Julie, 43–44

www.ingramcontent.com/pod-product-compliance
Lightning Source LLC
Chambersburg PA
CBHW031148020426
42333CB00013B/565